Rescue *in the* Valley *of the* Tigers

A Green Beret's inspiring story of daring,
courage, and selflessness — Vietnam 1968

Rescue *in the* Valley *of the* Tigers

A Green Beret's inspiring story of daring,
courage, and selflessness — Vietnam 1968

By

THOMAS A. ROSS

AMERICAN
HERITAGE
PUBLISHING

Atlanta, Georgia

American Heritage Publishing
Atlanta, Georgia

Rescue in the Valley of the Tigers

Published by
American Heritage Publishing
5710 Mt Repose Lane NW
Peachtree Corners, GA 30092-1428

First Edition - Privileges of War, 2004
Published by American Heritage Publishing Atlanta
5710 Mt. Repose Lane, NW
Peachtree Corners, Georgia 30092-1428

Second Edition – Rescue in the Valley of the Tigers, 2020

Book design by Jill Dible – Tom Ross
Editing by Robert O. Babcock & Robert L. Hawley, Jr.
Cover design by Creativindie Book Cover Designs

For legal and other reasons, some names in this book have been changed.

Library of Congress Control Number: 2020923942

ISBN: 978-0-9754859-3-4

Printed in the United States of America

DEDICATION

*To the more than three million who served, and, especially,
to each of the more than fifty-eight thousand who died or
were declared missing in action in the Republic of South Vietnam.*

*To my teammates on Special Forces Detachment A-502
and those encountered while there—you inspired this book.*

*To those who wear an American military uniform today.
Thank you for your service.*

*To the memory of my friend, Ngoc,
to his wife, Kim, and their four children; Chau, Bau, Hieu, and Nga.*

And, finally . . . with love,

*To Amy, my inspiration and—the light of my life,
my four children; Angie, Brian, Allison, and Lindsay,
my grandchildren, and my friends,
who all give me many good reasons—to live and enjoy life.*

CONTENTS

FOREWORD

THIS IS A REAL STORY. It is told by a real soldier about one of the most incredible moments in the Vietnam War. It is a story of the energetic innocence of youth, the craving to be a patriot, the anguish of war fought on the other side of the world. Yet, even in the despair amongst demons of horrific proportions, heroes were bred.

On one mission of mercy to save others, Tom Ross shows us what it means to be a positive American. He makes us all proud. He doesn't ask that you understand all the real pain of triumph and fear that he experienced. He only hopes that you won't forget what happened, or the men and women to whom it happened.

Tom Ross's *Rescue in the Valley of the Tigers* makes me glad to have served in Vietnam and proud to be an American.

Max Cleland
CPT U.S. Army, Ret.
Author, *Strong at the Broken Places*
Former U.S. Senator (1997–2003)

PREFACE

RESCUE IN THE VALLEY OF THE TIGERS is the companion book to *Along the Way*. So, if you have already read that book, you will discover that this one begins with a bit of the same familiarizing information.

As a young boy, I was healthy, very happy, and blessed with more than my share of vim and vigor. As a result, I gave my parents a run for their money as they did their best to raise me to be centered and have good moral values. To direct my energy constructively, my parents got me involved in Scouting at an early age. I became a Cub Scout and later a Boy Scout, ultimately earning the rank of Eagle Scout. My parents also made sure that I was in church every Sunday, where I served as an altar boy and choir boy.

My rearing took place in Pensacola, Florida, which was also home to the Navy's flight demonstration team, the Blue Angels. I used to watch them practicing out over the Gulf of Mexico while I fished from the Pensacola Beach Pier. They were magnificent and inspiring, so it was easy for me to grow up a patriot. As a boy, I always thought of myself as a good kid, an All-American kid. No one special, just someone who loved the country where he was growing up.

Later, during what had been carefree college days, I would find myself unexpectedly called to demonstrate my patriotism. The call was so strong that I would be compelled to leave college behind, join the U.S. Army, and volunteer for service in Southeast Asia. After nearly two years of exceptional training, I became a part of the Vietnam War as a military adviser. During my time as an adviser, I was privileged to meet and work with individuals who didn't hesitate to risk their lives for things they believed. Not all were military and not all had combat assignments, but they shared a common bond—all of them were Americans.

It may be hard to believe and difficult to understand, but my time in Vietnam was an incredible adventure. While it was full of both good and bad experiences, it was also full of challenges that tested my ability to respond. Any of those challenges could have abruptly ended my existence on God's green earth, so I came home with a very special appreciation for life.

This book is not a macho, shoot'em-up, war tale. It is a sensitive factual story of the inspirational dedication displayed by those who served our country during an unpopular war. And it was written so that it could be read and easily understood by women and other family members of veterans who may not have military experience. Finally, the book was also written for those born long after the Vietnam era so they may have another insight to compare with things they may have heard or read or were taught in school.

When I made the decision to enlist in the military and volunteer for service in Southeast Asia, I wanted my time there to be about more than simply fighting and surviving the war. So, when I left home, besides my duffle bag, I carried the hope of doing something good, achieving something meaningful while serving in Vietnam. Even though I had no idea what that might be, the hope still went with me.

In *Along the Way,* I mentioned my hope of accomplishing

something meaningful more than once. But, as the second half of my tour of duty began, it seemed unlikely that anything I could consider "meaningful" would occur. Then, remarkably, when I had nearly given up on the possibility, a unique mission would unexpectedly present itself.

The unusual mission, which I have always considered a gift, involved the rescue of families held captive and used as slaves deep within enemy territory. The unique mission received national recognition when the inhabitants of a remote mountain village were freed from a life of abuse and slavery.

Rescue in the Valley of the Tigers tells a more in-depth story than the one first told in *Privileges of War*, the first edition of this book. You will learn how the rescue began along with more interesting details about the steps required to locate the village and achieve the rescue—all in an extremely short period of time.

The rescue demonstrates the best our country has to offer those in need of help—anywhere in the world. I consider myself extremely fortunate to have been a part of that mission.

INTRODUCTION

RESCUE IN THE VALLEY OF THE TIGERS is the second of two companion books that represent the 2nd Edition of my book *Privileges of War.*

When I completed the manuscript for the second edition of the first book, the word count and page count were so high that the decision was made to divide the manuscript into two separate books.

Along the Way is the first of those two books and it describes selfless and courageous deeds performed by some of the individuals Tom met while serving his tour of duty in South Vietnam.

This is *Rescue in the Valley of the Tigers,* the second book and it describes Tom's long-held hope to do something good during his time in Vietnam and what happens when that opportunity presents itself.

WHILE SERVING AS AN officer and military adviser with the U.S. Army 5th Special Forces, it was my honor, my privilege, to witness American men and women in action.

In early February 1968, I was assigned as the Intelligence Officer (S2) and Operations Officer (S3) of Detachment A-502, the largest "A" team ever formed. Normally, an A-Team has twelve men assigned. However, because of its unique and important mission, 502 grew to a much greater strength and became shepherd to a very large area of operation.

I won't describe them here, but 5th Group Headquarters had given A-502's important mission requirements, which would fill several pages. It will be enough to say that's the team's key responsibilities were the defense of the 5th Special Forces Headquarters in Nha Trang, the Nha Trang Air Base, greater Nha Trang City, and the northern approach to Cam Rahn Bay.

As 502's S2 and S3, the team's broad responsibilities gave me the opportunity to work with virtually every branch of the U.S, military as well as many unique and specialized units. I came to refer to this powerful collection of support units as my "equalizers." If a situation ever became unbalanced against us, I simply called for support from one or more of the equalizers.

On more than one occasion, the equalizers either got A-502 team members out of death threatening situations or outright saved their lives, mine among them. Often, they came down through clouds in the sky, the same place where angels are said to dwell. And, there were those among us who thought of them as angels. It amused me that those big, dark green, noisy, often fire-belching machines, could be thought of as angels.

In the *Preface*, I mention that I left home with the hope of accomplishing something meaningful during the Vietnam War. When I was finally given that opportunity, it wasn't something I could do alone. Many others would have to rally and support me to ensure that the mission was successful. The story of the rescue that took place in the Valley of the Tigers is inspirational.

Ultimately, I hope that what you read will make you proud of all those who served our country with honor while in Vietnam—as well as all those who serve today in other dangerous places around the world.

If you enjoy this book, I hope you will consider reading *Along the Way*. It tells many different stories about some of our country's best and most dedicated men and women.

Author's Note

EVEN THOUGH INCLUDED IN my companion book, *Along the Way*, it is important to again share information about a unique relationship that developed between my Vietnamese Special Forces counterpart, Major Nguyen Quang Ngoc, and me. Major Ngoc commanded the Vietnamese troops my team and I advised.

The original draft of this book, *Privileges of War*, contained much more about a unique relationship that developed between me and Major Ngoc Nguyen.

When the book was ready for publication in 2003, I wanted Ngoc to review draft pages where he was mentioned to make certain he was comfortable with what I had written. Regrettably, with the multiple family and business moves we had both made, we lost contact. The last I heard from him, Ngoc was in Hawaii. I learned that when a coconut with my address and his return address carved on it was delivered to my front door. That was it—just the coconut.

Ngoc was an extremely intelligent man and at the time Vietnam fell, he had been promoted to colonel and was serving as the head of what amounted to the Vietnamese Veteran's Administration. However, he also had a warm sense of humor. As you will read, we often shared witticisms during our time together. Often, it was a way to lessen tension or divert from the distasteful aspects of war.

After removing the outer husk, I bored a hole in the coconut shell and drained the coconut milk. Next, I cracked the shell and harvested the coconut meat. Then, I gathered my family around our kitchen table and introduced them to the joy of fresh Hawaiian coconut.

Despite a diligent effort, which included contacting Vietnamese organizations in Washington, DC, and other parts of the country, I was unable to locate Ngoc or any of his children.

Years earlier, in 1975, and for reasons you will read later, Ngoc and I were sharing a bottle of wine at my Gulf Breeze, Florida home. That evening, he confided to me what seemed a deeply emotional hope that he could one day take his family and—return to Vietnam.

So, believing there was a real possibility that Ngoc may have returned to what had become a Communist Vietnam, I feared the repercussions he and his family might suffer if Vietnamese authorities read what I had written about his fight against them. So, to protect my friend and his family, I went back through the manuscript, changed occurrences involving Ngoc, removed all but the most innocuous references to him, then published the book in 2004.

Unexpectedly, in November 2015, Ngoc's daughter, Hieu, located me and called my office. I was extremely excited to receive her call. It was a wonderful moment that reunited me with the Nguyen family. And, as it turned out, Ngoc hadn't gone back to Vietnam but had still been living with his lovely wife, Kim, in Hawaii. While I was sorry that my many calls to "Nguyen" listings in the Hawaiian phonebook had somehow missed Ngoc, it was good to know that he had decided to stay in the USA.

Now, with this book, I can more accurately describe how important the man who became my friend was in making it possible for me to mount the daring—rescue in the Valley of the Tigers

This edition includes more stories, general information, and detail than the first. And, since they help tell a story, there are many more pictures of individuals and places you will read about. So, in many chapters, there are pictures of individuals and places you will read about—I want you to feel that you are there with me as you read each story.

ABBREVIATIONS

AHC	Assault Helicopter Company
AIT	Advanced Infantry Training
AO	Area of Operation
ARVN	Army of the Republic of South Vietnam
ASAP	As Soon As Possible
CAV	Cavalry
CIB	Combat Infantryman's Badge
Chopper	Helicopter
CIDG	Civilian Irregular Defense Group
CO	Commanding Officer
COC	Combat Orientation Class
DI	Drill Instructor
FAC	Forward Air Controller
FDC	Fire Direction Center
GI	Government Issued/General Issue
HE	High Explosive
KIA	Killed in Action
LLDB	Luc Luong Dac Biet (Vietnamese Special Forces)
LRP	Long-Range Patrol
LT	Lieutenant (Pronounced "ell-tee" in conversation)
LZ	Landing Zone
MEDCAPS	Medical Civic Action Program
MIA	Missing in Action
NCO	Non-Commissioned Officer
NVA	North Vietnamese Army

ABBREVIATIONS

OCS	Officer Candidate School
PIO	Public Information Officer
Psy Ops	Psychological Operations
ROK	Republic of Korea
RPG	Rocket Propelled Grenade
S2	U.S. Intelligence
S3	U.S. Operations
SF	Special Forces
SFC	Sergeant First Class
SFG	Special Forces Group
Sit Rep	Situation Report
SPC4	Specialist fourth class
Spooky	Modified C-47 aircraft gunship
SRAO	Supplemental Recreational Activities Overseas
TAOI	Team Area of Influence
TAOR	Team Area of Responsibility
UH-1	Huey helicopter
VC	Viet Cong
WIA	Wounded in Action
XO	Executive Officer

THE SETTING & FLAGS

North and South Vietnam (now, Vietnam) were both parts of the area known as "Southeast Asia." Surprisingly, many people weren't and still aren't sure exactly where the area is located. Southeast Asia is situated on the Asian continent and consists of countries located east of India, south of China, west of New Guinea, and north of Australia.

This map is provided so you will know where key cities and my base camp, Special Forces Detachment A-502, and Trai Trung Dung, were located within the overall arena of the Vietnam War.

THE SETTING & FLAGS

FLAGS

ARVN   Army of the Republic of South Vietnam

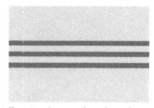

(Yellow Background with Red Stripes)

NVA   North Vietnamese Army

(Red Background with Yellow Star)

VC   Viet Cong

(Red Top, Blue Bottom Background with Yellow Star)

CHAPTER 1

Tell Your Story

IN LIFE, YOU NEVER know where inspiration, encouragement, or opportunity might spring from. Learning that at an early age, I have always tried to be prepared to take advantage of any of them—if and when they occur.

Once, I was fortunate to have all three walk up to me and shake my hand. Opportunity presented itself in the form of Robin Moore, author of "The Green Berets." I was given the opportunity to meet Robin during an annual reunion of the 5th Special Forces Group in Fort Campbell, Kentucky.

I had been asked to attend the 2003 reunion by Lisa Phalen, the wife of my friend and fellow A-502 teammate, Bill Phalen. I was to be there as a surprise for Bill, we hadn't seen each other since serving together nearly thirty-five years earlier. And, we had only recently reconnected.

If I went, that would be the first military social event I had attended since leaving Vietnam. So, looking forward to seeing Bill, I accepted Lisa's invitation and it was fortuitous that I did.

A few weeks later, my friend and I were reunited. And, as is often the case with veterans, it was a warm reunion and seemed as if not a single day had passed since we were last together.

Later, we were reliving, and Lisa was tolerating, some of our Vietnam experiences while sharing adult beverages in the "Hospitality Room." We had been talking for a while when Robin Moore, who was on a cane, and his entourage walked into the room. I didn't know who he was, but Bill said to Lisa, "Oh, there's Robin."

"Robin who?" I asked.

"Robin Moore. He wrote the book *The Green Berets.*"

"You're kidding! Of course, I recognize that name." I exclaimed, "I've written a book about our experiences."

"Would you like to meet him?"

"Absolutely!"

"Come on, let's go before he gathers a crowd," Bill said and we were up and on our way.

Bill moved so quickly that Robin and his entourage had barely cleared the doorway when Bill greeted him, "Robin, it's good to see you."

After they shared some small talk, Bill said, "Robin, I'd like you to meet my friend and former teammate, Tom Ross. He's written a book."

Robin was very kind and rather than seeming set-upon or uninterested, he asked, "What's it about?"

I told him that it was a good story about service in Vietnam and explained that I was having an extremely difficult time finding anyone interested in publishing such a book. I was shocked at what happened next.

Robin turned to his entourage and told them to go find seats, he would join them later saying, "I need to spend some time with Tom." Then, he directed me to an unoccupied corner of the room.

Robin's background is extremely interesting. As I've said, he wrote "The Green Berets," a bestselling novel in 1965, later turned into a movie of the same name, starring John Wayne. Robin also wrote the lyrics to Barry Sadler's 1966 number-one hit song, "The Ballad of the Green Berets."

Then, and even more successful, in 1969, Robin wrote *The French Connection* a nonfiction book that became an Oscar-winning film in 1971. That movie starred Gene Hackman as "Popeye Doyle."

Later, in 1972 he collaborated on *The Happy Hooker*, an international bestselling book. Unfortunately, the film adaptation in 1975, didn't fare as well as *The French Connection.*

Robin was able to write *The Green Berets* with such authority due, in part, to his connections in Boston. He and Bobby Kennedy were classmates at Harvard. And, Bobby's brother was, of course, President John F. Kennedy.

Modern-day Special Forces began to gain prominence in October 1961 when JFK visited Fort Bragg and the U.S. Army Special Warfare Center, then and still, the home of Army Special Forces. While at Fort Bragg, President Kennedy noticed that all of the Special Forces soldiers were wearing a green beret and he liked the look. At the time, the distinctive headgear was worn unofficially. Impressed by the demonstration of their capabilities during his visit, the President would later authorize the green beret as the "official" headgear of the U.S. Army Special Forces. From that point forward, the unit and its members would become forever known worldwide as—*The Green Berets.*

Pleased with what he had seen at Fort Bragg, the President further demonstrated his support for Special Forces and its unconventional warfare program. On, April 11, 1962, the President published an official White House Memorandum to the U.S. Department of the Army. In part, the memorandum read, "The Green Beret is again becoming a symbol of excellence, a badge of courage, a mark of distinction in the fight for freedom."

Because of the relationship that developed between President Kennedy and Special Forces, the U.S. Army Special Warfare Center was renamed the John F. Kennedy Special Warfare Center.

So, because of the close relationship between the President and General William Yarborough, the commander of Special Forces, Bobby interceded on Robin's behalf. He asked his bother if he would arrange for an introduction for Robin to General Yarborough. JFK agreed and an introduction was made.

Robin wanted to see the "Green Berets" in action at Fort Bragg, up close and in person. But, before General William Yarborough gave Robin full access to the unit, he insisted that he go through the Qualification Course ("Q course") training. The General wanted Robin to understand exactly what made the Special Forces "special." So, Robin agreed and went through the training.

As a result of his relationship with the unit, Robin became a lifelong champion of Special Forces. He even moved near Fort Campbell, Kentucky, the home of the 5th Special Forces Group. He had hopes of writing another book about the "Green Berets" and wanted to be near them. So, attending the reunion was easy for him.

When we reached the table, Robin pulled out a chair, sat, then hung his cane on the back of the chair next to him. "Can you give me a brief synopsis of your book?" he asked.

"Of course," I said. Then, I explained that the book was divided into two parts. I told him that Part I was titled "Along the Way" and discussed those I met during my tour of duty, the non-combat ones you remember and the ones met in combat who provide support and help keep you alive.

"Yes, I know about them, Robin said, "I learned about them firsthand when I was attached to the 5th Group in Vietnam."

That's when he told me that General Yarborough insisted that, if he were going to write about them, he needed to become one of them. And, that's what he did.

Then, after completing training, he said that President Kennedy personally authorized his deployment and attachment to the 5th Group.

It was after his service in Vietnam that the men of 5th Group had more-or-less adopted him, which is why he was invited to every reunion and convention, and how my friends, Bill and Lisa Phalen, got to know Robin.

Then, "Okay, tell me about Part II," Robin said.

Responding, I told him that Part II discussed a rescue mission I led that freed a remote mountain village from slavery and abuse.

He seemed genuinely intrigued and said, "That's what you guys do . . . tell me more."

For about the next thirty minutes, I provided more detail about the rescue and Robin asked questions. Then, I shared my frustration with not being able to find a publisher interested in publishing a positive story about service in Vietnam. After discussing that issue for a while, our conversation drifted a bit to things still military, but more personal.

We talked about things we had seen in Vietnam and, after he told me that he had served during WWII and had been the nose-gunner in a B-17, I told him that my Dad had also flown in B-17s during the same war and his plane had been shot down.

Our visit became more lighthearted when Robin talked about his meeting "the Duke" (John Wayne) during the filming of the movie version of his book about the Special Forces.

Eventually, we returned to my book. Robin looked at me and was quiet for a moment, then he unloaded, "My book (*The Green Berets*) is fiction. Yours is nonfiction, it tells real stories about real people. And, the rescue tells a wonderful story about what Special Forces is about. Tom, forget other publishers. Go back to Atlanta and publish your book yourself . . . I published my first book. Tell your story!"

And, with that as guidance, that's exactly what I did. *Privileges of War* was published in 2004, approximately one year after my encounter with Robin Moore, a very kind man who was generously willing to share his time and offer encouragement to a first-time writer. When we got up to leave the table, we had spent nearly an hour together.

Despite the briefness of our encounter, I felt a loss when I heard the announcement of Robin's death on February 21, 2008. Even though he is gone, I will not forget him because of the inspiration he left with me. And, that inspiration was, in part, responsible for this book.

CHAPTER 2

Enough Chances

BY THE FIRST OF August, 1968, I had been serving as the S2 and S3 officer of 5th Special Forces Detachment A-502 in Vietnam for about six months. Our basecamp, Trai Trung Dung, was located approximately twelve "klicks" (kilometers) west of the beautiful seaside city of Nha Trang, a place that has become a much-traveled-to tourist destination since the war. The beaches in Nha Trang traced the edges of the crystal-clear blue waters of the South China Sea and were almost as beautiful as the white-as-snow beaches in my hometown of Pensacola, Florida. Flying along the Southeast Asian coastline on the day I first arrived in Nha Trang, I imagined the setting as a tropical vacation destination.

A-502 reported directly to the 5th Special Forces Group Headquarters, which was located in Nha Trang City, not far from the beach. For the non-military reader, Special Forces was and still is the elite military unit also known as the "Green Berets." 5th Group provided direction for all SF (Special Forces) operations in the Vietnam theater of war. However, rather than beach towels and suntan lotion, they had given the men of A- 502 a huge and far more serious job with important responsibilities.

Originally, A-502's mission was the defense of the 5th Group Headquarters and the Nha Trang Air Installation. That was a big enough job. But later, when it was determined that the defensive perimeter needed to include the greater Nha Trang City area, the responsibility for that assignment was added to A-502's mission. Later, because of its strategic location in the Nha Trang Valley, 502 also assumed responsibility for the defense of the northern approach to Cam Rahn Bay.

At its greatest strength, which was the year I was there, A-502 had fifty-five American military advisers assigned and we worked out of our Trai Trung Dung basecamp. Trung Dung was located in an old Vietnamese fort, known to locals as "The Citadel." The fort had been occupied by the French when they were in Vietnam and it looked like a Hollywood movie set to me. I kept expecting to run into Cary Grant or Humphrey Bogart in an old French Foreign Legion uniform.

One night, while up on the fort wall, I encountered a Vietnamese soldier on sentry duty. He spoke English well enough that we could carry on a conversation. When I asked him how old the fort was, he said, "I not sure, but so old . . . ghosts live here."

Later that evening, when I mentioned the soldier's comment to a teammate, he said, "Yes, several of the Vietnamese say they have seen ghosts around the fort."

Many strange things happened in Vietnam and, believe it or not, ghosts weren't the strangest. In my role as A-502's Intelligence Officer, I was at 5th Group Headquarters one day in 1968 when I learned something unbelievable. As if fighting the NVA and VC weren't enough, I was told that a UFO had attacked an American patrol boat in the DMZ (Demilitarized Zone), the area that separated North and South Vietnam. It seemed two U.S. patrol boats had a close encounter while on routine patrol one night. But ghosts and UFOs are other stories and, maybe, another book.

Officially, A-502's mission was stated as follows:

General:

(1) Advise, assist, train, and support the VNSF (Vietnamese Special Forces) in their CIDG (Civilian Irregular Defense Group) program. These were essentially citizen soldiers.

(2) Conduct position defense, defend the SFOB (Special Forces Operational Base) Headquarters from enemy attack and be prepared to conduct counterattacks on order.

Specific: These specific missions required A-502 to produce an OPLAN (Operations Plan), an Operational Plan describing exactly how they would carry out each mission. Keeping this important plan up to date was one of my jobs.

(1) Prepare defense of A-502 and all of its Outposts.

(2) Prepare to assist in the defense of the Nha Trang area and occupy key terrain and blocking positions on order.

(3) Provide early warning through Binh Tan and Thuy Tu outposts.

(4) Provide supporting fires in the sector.

(5) Construct defensive positions to include overhead cover, when possible for all personnel in the task force sector.

(6) Develop defense plans and counterattack plans for the assigned sector.

I'm not going to bore either the civilian or knowledgeable military reader with more mission details. I will simply say that, in addition to "General" and "Specific" mission responsibilities, there were also many "Implied" responsibilities. I share this information simply to demonstrate that A-502's overall mission assignment was a huge responsibility for such a small group of men. However, it was one that the men of U.S. Army Special Forces Detachment A-502 eagerly accepted.

With such sweeping responsibilities, my dual jobs as the team's S2 and S3 officer were both interesting and challenging.

The jobs were interesting because every day was different as my Intelligence Sergeant, Paul Koch, and I tried to stay ahead of the enemy, trying to determine where they would pop up next. Then, when they did appear or when we went looking for them, my Operations Sergeant, Roy King, and I would have to be prepared to suggest the best strategy for A-502 to mount a response.

The jobs were challenging because the VC (Viet Cong) and the NVA (North Vietnamese Army) had more than enough chances to kill me. I was in the field constantly, visiting our outposts or flying various observation missions over every square meter of our AO—and beyond.

Our intelligence told us that the NVA routinely approached Nha Trang from the west. For that reason, some of my observation flights with both Army and Air Force fixed-wing pilots took us far beyond the western border of 502's area of operation in the direction of Cambodia. We were in search of trails and, if we got lucky, an enemy unit on the move in the open. If that happened, we would call in an airstrike.

As we flew our searches, some of the pilots seemed qualified to join a flying circus. In an effort to locate enemy encampments near rivers, the pilots would dive down into river bottoms or just above them and contour fly the river's path. The hope was that by flying low (very low), our observation plane wouldn't be detected before we sighted the enemy unit.

About these missions, I must confess two things. They were very exciting and I almost always became airsick. But, for the pilots who flew these observation missions daily, they were also very dangerous. Flying so low exposed them to enemy ground fire. One of the Air Force pilots who worked in the area flew out to the west one day—and never returned. We never learned his fate.

And, that's the way it was for every member of our team. Each mission planned for and undertaken by members of the A-502 team put their lives at risk. For that reason, I took my responsibilities seriously and was very much into my job. I knew being out in the field so much was dangerous, but that wasn't something I thought about daily. What I did think about was that it was my job to know as much as I could about the enemy. So, I did the best I could to be informed and stay alive.

To keep me alive, Sergeant Koch began cautioning me about leaving camp alone to visit the village of Dien Khanh that surrounded the fort. He told me that one of our intelligence agents and the Vietnamese Intelligence team had informed him that a bounty had been offered for my life. The reason was that because A-502 ambush units, manned by combat-experienced Vietnamese soldiers and advised by American team members, had a very high kill ratio. While I wasn't always the one who pulled the trigger, the enemy needed someone to blame. So, they chose the American Intelligence Officer at Camp Trung Dung as the offender—me.

The first time Sergeant cook told me about the bounty, I thought he was just playing with my mind. But, after dismissing what he said and laughing at him, he became upset with me and yelled, "No bullshit, sir! They'll kill you if they get the f---ing chance!" At least he said, "Sir."

Assuming the information was accurate and the threat real, it wasn't something I could worry about every day. I still needed to keep my focus and do my job. But, behaving prudently, I simply quit going to Dien Khanh and never left camp without another American or one of our Montagnard bodyguards. Occasionally, however, I did taunt my friend and teammate, Sergeant Koch, by telling him that I would give him a hundred U.S. dollars if he could get me a "Wanted" poster for a souvenir. Strange, but making light of the situation made it easier for me to focus my attention elsewhere.

The danger of assassination was very real for every American adviser at A-502 and everyone knew that to be true. Any local Viet Cong would gladly take an unexpected opportunity to kill an American. So, everyone made sure to have someone else with them when they left camp. Such was just one of the many day-to-day precautions taken to give each of us the chance to return home alive and, hopefully, uninjured.

During my time at A-502, my job was made much easier by Major Nguyen Quang Ngoc, the Vietnamese Camp Commander. Our team served as advisers to him and his Vietnamese troops, which included five companies of VNSF and CIDG soldiers.

On the first day that I met Major Ngoc, he and Major Lee, A-502's CO (Commanding Officer), were standing just outside the team-house when I was dropped off to reported for duty.

When I first met Ngoc Nguyen, I could easily have mistaken him to be an accountant. He was mild in tone, gentle in manner, and warm in personality. Other than his combat attire, there was no suggestion of the combat seasoned soldier who had led a fierce counter-attack on the North Vietnamese Army and Viet Cong during the last few days.

After greetings and introductions, Lee explained to Ngoc the role I would play at 502. He said, "Lieutenant Ross will be our Operations and Intelligence Officer. Give him a day or two to come up to speed, but anything you need from A-502 going forward, he's your man."

Ngoc welcomed me to Trung Dung and after some small talk regarding a gunfight my driver and I had encountered on our way to 502, I made the statement that I was glad to have reached Trung Dung safely and had survived the day. The words had barely crossed my lips when Ngoc smiled and, waving his index finger, he said, "Day not over yet Trung uy."

Trung uy (pronounced, Trung wee) is Vietnamese for the rank of 1st Lieutenant and Thieu ta (pronounced, To ta) means Major. And, that was how we addressed each other for nearly a year, I called Ngoc, Thieu ta, and he called me, Trung uy.

After, cautioning me that the day was not yet over, Ngoc asked Major Lee if he would assign me to Buddha Hill that night to serve as one of his advisers.

At that moment, I had been in Nha Trang for less than four or five hours, so I'm not sure how effectively or intelligently I could have advised him. I suspected that he simply wanted to get to know the person who he had just been told would supply him with things he needed. So, when Major Lee agreed, I just went with the flow.

I had arrived in Vietnam on the heels of the 1968 Tet Offensive, a major surprise attack mounted by almost seventy thousand North Vietnamese and Viet Cong soldiers. The attack, launched during Tet, the lunar New Year, violated a holiday truce and occurred virtually simultaneously all over the country. In a single night, the war moved from the jungle and rural villages to the heart of over a hundred Vietnamese cities and towns, some previously thought to be impregnable. Nha Trang had been one of the first coastal cities hit as its inhabitants prepared to celebrate the New Year, the Year of the Monkey.

According to superstition, the monkey is considered a harbinger of bad luck. Certainly, in this case, it was. During the Tet Offensive, South Vietnam and its allies lost thousands of troops, hundreds of American soldiers among them. However, the Viet Cong and North Vietnamese who launched the attack lost thousands more.

While dissension had been growing regarding the war in Vietnam, the Tet Offensive would dramatically change public opinion in the United States and around the world. Events surrounding this offensive would cause antiwar resistance to intensify significantly.

The NVA and Viet Cong were still active in the Nha Trang area and no one knew how long the Tet Offensive would continue. So, Ngoc wanted to ensure that the enemy didn't attempt to retake Buddha Hill as they had the first night of the attack. That's why we were going there and where I would spend my first night in a combat zone. And, it would be on Buddha Hill that Ngoc and I would begin to build a lifetime friendship.

As my tour of duty unfolded, Ngoc and I would share many experiences and I would become a seasoned military adviser. During that time, I would also have the opportunity to meet and work with some exceptional and courageous American men and women.

Experiences with Ngoc, my seasoning, and my introduction to those I met and worked with are the subjects of *Along the Way*. Now, beginning where that book ended, you will learn how the rescue in the Valley of the Tigers began and how it became national news. And, while the NVA would have yet another chance to claim my life, others would intervene to ensure they weren't successful.

Entrance to 5th Special Forces Group Headquarters

My assignment with the 5th Group would lead to one of the most rewarding experiences of my life, an experience I will never forget.

A-502's Area of Operation & Outposts

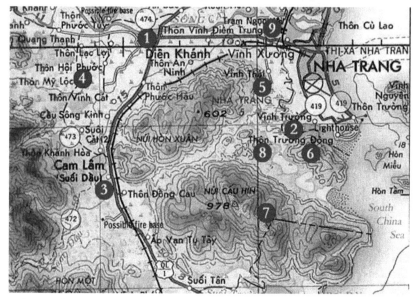

This is the heart of A-502's AO (Area of Operation). This is where team members spent most of their time working and risking their lives daily. Long-range reconnaissance patrols, or LRRPs (pronounced "Lurps), would often take team advisers far to the west (well off this map) in search of the enemy. Below, is a list of our camp and outposts.

1) Camp Trai Trung Dung – Base Camp

2) Binh Tan

3) Soui Dau

4) Nui Thi

5) Thuy Tu

6) Buddha Hill

7) My Loc

8) OP Ngoc

9) Da Hong

CHAPTER 3

Nha Trang

WHEN FIRST ARRIVING IN Nha Trang, I found it difficult to believe that there was a war in such a beautiful place, but it didn't take long to convince me otherwise.

In this second edition of *Privileges of War*, there are many more pictures throughout the book so that you can see much of what you will read about. You will see Vietnam as I first encountered it, where I worked, and witness the rescue as it unfolded. Fortunately, the event was extremely well documented.

Postcard View of Nha Trang

This was my view the day I arrived in Nha Trang. The ride brought me up the coast and around that point in the distance. The beach was nearly as white as the one back home—Could there truly be a war here?

Circling to Land in Nha Trang

The crew chief's .60-caliber machine gun in this picture suggests that the answer to my question was—yes. But that was difficult to believe.

Since the war in Vietnam, Nha Trang has become a popular travel destination. One of my friends is a travel agent and takes tours there. When she asked if I would like to join her tour, I said, "I've been there."

Special Forces Headquarters

5th Special Forces Group Headquarters—Nha Trang, South Vietnam. 5th Group was responsible for all Special Forces operations in Vietnam.

Welcome to Southeast Asia

This colorfully painted temple wasn't far from 5th Group Headquarters and was evidence that I was in a very different part of the world.

Buddha — Nha Trang Icon

Buddha, an icon of Nha Trang, sits atop a hill and watches over the city. This was my view of the huge statue as I was driven away from 5th Group Headquarters, through the city, and out through the Nha Trang Valley on the way to my new home at Detachment A-502.

Buddha Hill

My first night as the S2 and S3 of A-502 would be spent on this hill to prevent the enemy from retaking the tactically important location. The story of what happened that night is a chapter in "Along the Way."

Buddha had two descending dragons guarding his steps and they looked very different at night. The hill took on a surreal appearance.

Driving Through Nha Trang — Street Scenes

Nha Trang was the first city attacked during Tet. Even though the attack was only days earlier, the streets initially seemed calm. That impression was deceptive because they had already exploded on me.

Street Vendors and Local Residents

As we made our way through the city on our way to A-502, the air was filled with smells of cooking food that I had never experienced before. They weren't offensive, just very different.

Family Bakery

Local residents, street vendors, and shopkeepers were still cleaning up from TET and trying to return to normal as best they could.

A Laurel and Hardy movie was playing at this theater. All of these pictures show a city that seemed to be going about business as usual. Then—Boom! The story of what happened to me in the streets before I reached 5th Group Headquarters is also a chapter in "Along the Way."

CHAPTER 4

Trai Trung Dung

TRUNG DUNG WAS MY base of operation and my home while I served in South Vietnam. The following pictures take you beyond Nha Trang to show you the old fort and A-502's location in it.

The fort was the second oldest such structure in Vietnam, so I found it very interesting. And, I was greatly amused that I worked out of a location that had also been used by the French Foreign Legion. I even found an old map that showed the fort, complete with its moat. You will see a piece of that map on the next page.

According to locals, the fort was built under the Nguyen Anh Dynasty around 1793. The Citadel was strategically located north of Cam Ranh Bay and west of Nha Trang, near the village of Dien Khanh, in Khan Hoa Province.

When first constructed, from its key location, those occupying the fort could help defend the seaport cities by patrolling the east to west Nha Trang Valley and the pass running north to south toward Cam Ranh Bay.

More recently, Seabees, a navy engineering unit, had built and occupied the camp prior to the A-Team moving in. 502 inherited the facilities when the Seabees moved to work at a new location. The unit left A-502 with accommodations that were far above average for an A-team encampment.

The first day I arrived at Trai Trung Dung and A-502, I had been driven through the Nha Trang Valley with its rich history dating back more than four thousand years. The following pictures will introduce you to Dien Khanh Village, the Citadel, and Trung Dung similar to the sequence I encountered them.

First, the Nha Trang Valley was a beautiful picturesque patchwork of villages, lush fields, and rice paddies that were framed by mountains on its north and south.

Nha Trang Valley

This is just a tiny portion of the Nha Trang Valley. It was always surprising to find beauty in a country filled will war. The natural beauty of Vietnam and the sometimes-horrific occurrences of the war seemed unfortunately contradictive.

You can see the narrow road that ran between Nha Trang and the village of Dien Kha in the upper left of the picture.

The Village of Dien Khanh

The narrow road in the last picture eventually led us into the village of Dien Khanh. When I had free time and nothing else to do, I would either take one of the jeeps or borrow a motorcycle and visit the village and its small shops. I came home with a few gifts for my family that I had purchased in Dien Khanh.

Local Restaurant

This was one of the Dien Khanh "open-air" restaurants that were frequented and recommended by Vietnamese soldiers.

Dien Khan Rapid Transit System

The East Gate of "The Citadel

The eastern entrance looked as if it were a part of a Hollywood movie set.

"The Citadel" — Old Vietnamese Fort

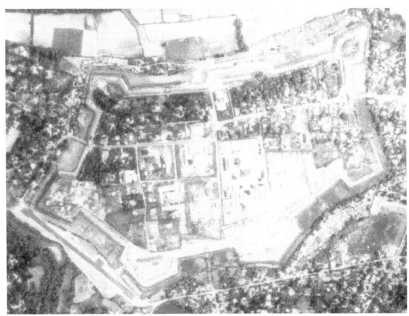

The Citadel was once home to the French Foreign Legion. Intelligence Photograph by the CIA (Central Intelligence Agency).

East to West View

This is the east to west view of the Citadel and its placement in relation to Dien Khanh Village which surrounded it.

Base Camp

This is the entrance to Special Forces Detachment A-502's base camp.

A-502's Main Compound

Our main compound is inside the pointed portion of the fort below the white dashed line in the picture above. You can see the gardened flag pole frame and flag in the center of the compound. Our area extended to the left to the runway.

East-to-west view and you can see across the compound to the runway. A-502 contracted to facilitate operations.

Old Map

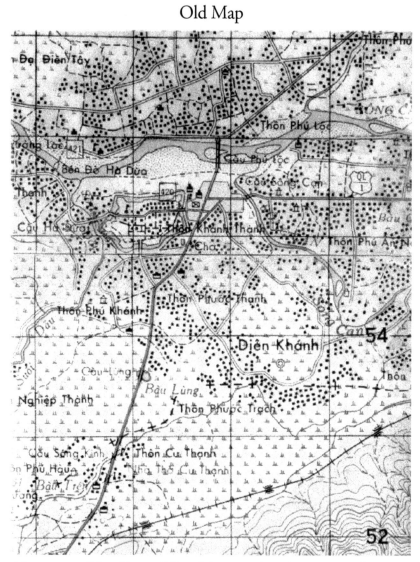

This is part of the old map I found. It certainly isn't ancient and I suspect it may have been one used by the French when they were in Vietnam.

CHAPTER 5

A Promise Made

IT WAS HAIRCUT TIME. My hair had grown almost three inches long on top, and I was sure Major Lee would say something if it grew much longer. Amusing to me, but Will Lee was an extremely disciplined man who believed an officer's hair should always be neat and closely trimmed, even in a combat zone. Mine was neat but long enough to comb. A trip to the barber was probably in order, especially if it kept my CO happy.

It was about 10 a.m. when I left the small two-chair barbershop operated by a couple of the camp's Vietnamese soldiers. I was still brushing hair off my neck when Sergeant Koch came walking up at a quick pace. He seemed in a hurry.

"Sir," he said, as he waved a salute.

"Good morning, Sarge. How does my haircut look?"

"Fine, sir. It looks really good."

So much for our exchange on my haircut. He obviously had something more important to talk about.

"LT, we just got a call from My Loc. They say three VC have surrendered to them. They want you out there ASAP," Koch said.

"Fine, find Ahat, grab a jeep, and we'll head that way."

Ahat (pronounced "Ah-ott") was my Montagnard radioman and trusted bodyguard.

The Special Forces often used Montagnards as radiomen, bodyguards, and trail watchers because of their loyalty and dependability. The Montagnards, who had also fought with the French when they were in Vietnam, were a tribal people who often lived in remote areas throughout the country. They kept to themselves and bothered no one.

Frequently referred to by Americans simply as "Yards," the Montagnards were looked down upon by the Vietnamese. A minority group, they were considered inferior because of their primitive lifestyle. Vietnamese called them *moi,* which meant savages.

Typically, two or more of our SF team members would go out on a mission serving as advisers to a squad, platoon, or company-size Vietnamese force. Unsettling, but our intelligence suggested that as much as 15 percent of the Vietnamese CIDG troops in camp could be VC sympathizers. So, the Montagnards were used in the dual role of radiomen and bodyguards. During the day and at night they took turns standing watch over their American advisers, if and when they slept.

Despite the primitive nature of the tribesmen, SF advisers and the Montagnards they trained developed many warm relationships as they shared some of their respective cultures with each other. It was not unusual for an SF adviser to be taken as an honorary member of the Rhade or one of the other Montagnard tribes. The induction ceremony could be elaborate, taking as long as two hours, or it could be as simple as giving the honoree a carved metal bracelet, which symbolically bound him to the tribe. Such a bracelet given to me remains one of my proudest possessions.

While they were primitive, the Montagnards seemed to me to be an attractive, gentle people with many fine qualities. They had dark-brown skin, straight black hair, and larger, rounder eyes than the Vietnamese. They looked somewhat Polynesian to me, though with darker skin.

The Montagnards and their consistently demonstrated loyalty to SF soldiers were critically important and meant a great deal to the men of A-502. One day, when desperately needed, team members would be given the opportunity to reciprocate. The team would launch a bold and daring rescue mission into enemy territory in an attempt to free the enslaved inhabitants of a Montagnard mountain village—while risking their own lives in the effort. This book describes that mission in detail.

While Koch went to find Ahat and a jeep, I went for my gear, and in a few minutes, we were on our way west to My Loc. On the way to the somewhat remote outpost, I queried Sergeant Koch about the call he had received. No enemy soldier had surrendered before. "So, what do you think, Sarge? Seems a little strange, doesn't it?"

"It sure does. Nothing like this has happened as long as I've been here," he said.

"Well, I guess we'll find out what's what soon enough."

Upon arriving at My Loc, we were taken to the main bunker where the three VC suspects were being held. Sergeants John Herbert, Gary Dubovick, and specialist James Everett were the three permanently assigned A-502 advisers at the outpost. It was one of them who placed the call to Sergeant Koch.

Immediately upon seeing the surrendering trio and after talking with them via an interpreter for just a few minutes, I had a much clearer understanding of what was happening.

It was August 2, 1968, and by that date, I had been in Vietnam for nearly eight months and had become a seasoned adviser. So, as soon as I saw the men, I knew they weren't VC at all. I immediately recognized them to be Montagnard tribesmen—and they had a story to tell.

All three men were small, had dark-brown skin and jet-black hair. They appeared frail and were barely covered in tattered khaki clothing. Mang Quang, who seemed to be the dominant of the three, claimed that the inhabitants of their village had been enslaved by the VC and NVA. He said this had gone on for years and explained that he and the other two men had escaped two days earlier. And, they hadn't made their way through the jungle to turn themselves in as VC. Their desperate two-day search for our outpost was a journey made for an extremely important purpose.

Two of the men found Chieu Hoi (pronounced "chew hoy") passes that had been dropped from allied aircraft into fields near their village. The small leaflets were part of a joint military Psy Ops (Psychological Operations) program. They promised a warm reception and good treatment to enemy soldiers who would quit fighting and turn themselves into allied forces. The hope was that VC or North Vietnamese Army soldiers would find them and surrender.

When the Montagnards found the passes, they thought that, if they could make their way to the allies and turn themselves in, they might be able to obtain help for their village.

Now that I knew who they really were and why they were there,

I asked the villagers to continue telling their story through Ahat, who now served as our interpreter. What followed was a tale of servitude and abuse. It seemed that the VC and NVA had made slaves of not just these three men, but the entire village as well. For the past eight years, the villagers had been forced to serve as pack animals, carrying ammunition and supplies to various locations throughout the area. They were also forced to grow food to be used by enemy units passing through the village. Their story of captivity included tales of brutal beatings and other more violent atrocities perpetrated on women and children. The stories were disgusting and made me very angry—I had always hated bullies!

Mang Quang was a village elder and claimed to have been kept restrained in the village. He explained through the interpreter that after years of abusive captivity, he had grown weary of the VC/NVA treatment and began to defy their orders. Rather than killing him when he became defiant, his captors told him that he would serve their purposes better as a living example for the rest of the village. They responded by beating him regularly and keeping him tied to a post at night for the better part of two years. He had rope burns and the other two men as witnesses to support his story.

Mang Quang said he recently convinced the VC of his contrition and asked to be released from his in-camp duties and allowed to work in the fields. He said he decided to make the plea after his two friends told him about the information in the Chieu Hoi passes they found.

Early in the morning after the VC released him and permitted him to return to the fields, he and his two friends quickly, but quietly, disappeared into the dense jungle. They hoped to find the allies and the things promised on the pieces of paper they carried with them.

Even though the three had been successful in their attempt to reach our outpost, Mang Quang was noticeably upset as his story continued to unfold. The reason for his distress was quick to follow and easily understood. He explained that, when they released him to work the fields, his captors gave him an extremely stern warning about checking in on a routine basis. They told him that he must check in with them every three days. They said that if they hadn't seen him in three days—they would kill his family, a wife, and two young children.

When he again began to speak, Mang Quang said the VC would think he slept in the fields one or two nights since that was a common practice and because he had just been released. But, if he didn't return by the third night, he felt sure they would know he had run away and would very likely carry out their threat to kill his family.

One of the other Montagnards quickly spoke up and said that the VC routinely threatened all of the villagers and had killed other families before. He said that was the reason they came to us. They wanted to "turn themselves in" because they needed help for their village.

Mang Quang then continued his story. When he finished, Ahat, translating the villagers' emotional pleas, turned to me and translated what had been said in English, "He says they are not able to defend themselves or their families from the VC, Trung Uy. They need help because they don't want their wives or children to be hurt anymore." Ahat's lips were quivering as he spoke the last words of Mang Quang's plea for help and he appeared visibly shaken by the story he had just translated. He wasn't the only one. Sergeant Koch and I were both moved, as were our American teammates who had gathered in the bunker.

In many cases, the Montagnards were very primitive; often they had only their traditional crossbows with which to hunt food and defend themselves.

Obviously, so poorly armed, they were no match for the VC or NVA, who either by force or intimidation would press the Montagnards into servitude. But to the Americans who clothed, fed, and offered medical care to them, the Montagnards were fiercely loyal. After training from SF advisers, they proved to be equally fierce soldiers.

Because Ahat had obviously been translating for me, Mang Quang turned to face me and looked directly into my eyes. And, his were filled with tears. Slowly, he reached out and took one of my hands with both of his. Then, in his language, as Ahat translated, he begged for help for his village. I can't adequately express to you how his pleas and the desperation in his voice made me feel as the man held my hand and asked for my help. I immediately imagined my own family in the horrible situation he had just described.

And, as if what he had just said weren't enough, through Ahat, Mang Quang reminded me that, tomorrow would be the third day. If he didn't return to the village with help by the next day, he believed his wife and two children would be killed.

As I thought about all that I had been told since arriving at the outpost, I suddenly felt the tremendous weight of what I had been asked to do.

At the time of my encounter with Mang Quang and the other two villagers—I was twenty-two years old.

As I continued to absorb and consider the villagers' request for help, I realized that I faced very serious dual responsibilities. I would be responsible for whatever might happen to Mang Quang's family and the other families in the village if we did not respond. At the same time, I would be equally responsible for what might happen to members of the rescue team if we did help the villagers and were met with disaster.

While I stood there in the bunker and considered options, I looked at the faces of my teammates. I looked at the Special Forces patch on the shoulder of Sergeant Koch's uniform, then I looked at the Special Forces crest on his green beret. It read, "De Oppresso Liber"—Latin for "Free the Oppressed." That's the Special Forces motto.

As I thought about some of the many possible results, it occurred to me that if you are wearing the uniform of an American serviceman or woman and you are asked for help, there is really only one possibility. If you have the means—you provide the help.

Mang Quang's gaze was still fixed on my face when I looked back into his eyes. By that time, Ahat had become accustomed to translating almost simultaneously. So, as soon as I began to speak to Mang Quang, he immediately began to translate. My response was simple and clear. Returning his fixed gaze, I spoke directly to Mang Quang, and said, "We will give you the help you need."

Upon hearing the news, Mang smiled broadly, nodded his head, then he again took my hand with both of his—and he began to cry.

Spontaneous responses of, "Yes!" "Yes!" were immediately voiced by my American teammates. I was glad to hear their approval and knew they felt as I did, we were going to have to help these men.

Every American present immediately volunteered to be a part of the rescue team. I can't explain how extremely proud that made me feel about the men with whom I was serving. Despite the many risks and unknown challenges that they knew we would face; they didn't hesitate to step forward. Their commitment made me absolutely certain that I had made the correct decision.

After regaining his composure and still squeezing my hand tightly, Mang Quang again spoke directly to me. Without understanding his words, I felt the message he was trying to communicate to me. My feeling was confirmed when Ahat translated Mang's words.

"He is giving you great thanks, Trung Uy, for helping him and his people."

Nodding, I returned his smile saying, "Tell him that I understood what he said. Then tell him that we are glad to help. That's why these other men and I have come here."

As a young idealistic American Special Forces officer, my words weren't concocted, they were sincere and quite natural—because that was exactly why I was there. I've said it before and will say it again, I hadn't gone to Vietnam to kill anyone. I was there to help a country in its fight for freedom. For me, my presence in Vietnam was that simple and I know the same was true for my teammates and others.

Looking at the three frightened villagers huddled together and giving thought to the situation, I realized that, while we were going to do something for them, they had already done something very important for us. They had given me and the other members of A-502 the opportunity to actually live the Special Forces motto—Free the Oppressed. There was no way that my team or I could have ignored their pleas for help.

As Americans, we are helping people. This statement is easily proven true by simply observing our actions whenever there is a disaster somewhere in our country or some other part of the world. We are quick to respond and among the first to send help. Very simply, our nature is to respond when others are in distress, and this is one of the qualities that define us as Americans—it's in our blood.

I had just made a huge promise that would involve a great deal of planning and many people. Now—I was going to begin doing my very best to keep it.

Meet Mang Quang

Mang is standing beside a standard military jeep and he isn't much taller than the hood of the jeep, but he had as much courage as any man I have ever met. This picture was taken the day we met. He had arrived in tattered khaki rags and he loved this mismatched outfit he was given.

Inspiration from a Patch and Crest

This is the patch and crest worn by Sargent Koch that had provided the inspiration to take on a mission that I knew could become dangerous for all those involved. And, it would—far more than expected.

My Loc Outpost Entrance

Sergeant Koch and I came and went through the arch and guard post at the bottom of the hill. This view is looking northeast. The outpost was ringed with several rows of barbed wire to slow down any attack.

My Loc Outpost

My Loc, a tiny, but strategically located outpost on top of a small hill.

Death Before Freedom

This Vietnamese gunner is behind My Loc's 50-caliber machine gun. Because the outpost was isolated, they had to be constantly vigilant. This gunner could have easily killed the three Montagnards when they crossed the outer barbed-wire perimeter. Fortunately, those who first saw them coming allowed the villagers to approach because they were unarmed.

My Loc's Main Bunker

This is the outpost Mang Quang found and where I first met him.

My Loc's Main Mortar

Those stationed at My Loc used this mortar emplacement to support many of A-502's night ambushes. If the Montagnards had appeared threatening, this weapon could easily have been used.

CHAPTER 6

Dealing with Bullies

AS YOU BEGIN THIS chapter, I ask that you remember that in my youth I was an Eagle Scout, an altar boy, and a choir boy. Again, no angel, but I was a good kid with a great zest for life and a belief in justice.

As a younger boy, I often carried a pocket full of rocks I had collected, and much like one of Mark Twain's characters, I could easily show you where the nearest frog could be found, or a creek where you could swim and fish. My friends and I could usually be located in nearby woods where we would likely be climbing trees, building forts, or constructing tree houses. And, my dog, Bingo, who was always with me, would very likely be running around barking at all of us. Those were great days and I had not a care in the world. And, I was never one to look for a fight, but I would certainly never run from one.

I have always hated loudmouths, jerks, and especially bullies. They come in many forms and have numerous and, often, serious psychological issues. And, I am qualified to make that statement, but that's another story, another book, and isn't the subject of this chapter. This very brief chapter is about how I once dealt with a bully and why I felt as strongly as I did about the way the Montagnards were being treated.

As soon as I learned that villagers were being abused and hurt, they had me in their corner as a supporter—instantly. Especially, when I learned that the abuse also included women and children. What kind of a man would abuse a woman or child?

With the above as an introduction, I will now tell you that I attended what, in my days, was referred to as junior high school at Little Flower Catholic School in Pensacola, Florida. I was in the 8th grade, which is now referred to in most places as *middle school.*

Early one morning, I rode my bicycle onto the schoolyard and found my friend, Pete, being beaten to a pulp by a schoolyard bully who towered over him. In fact, he also towered over me because I wasn't much taller than Pete.

Pete had been punched in the face, had blood smeared across cheeks and blood was dripping from his nose.

While I wasn't much bigger than my friend, it's amazing how quickly a situation can be made more evenly balanced by using an "equalizer." There are many types of equalizers. And, as you might imagine, in Vietnam, I had an array of equalizers from which to choose. Artillery of different sizes, air support of different types, and more. But the morning I decided to give my friend Pete a hand, a far less impressive equalizer just happened to be nearby. It was a piece of a broken tree branch about two and a half feet long and about two and a half inches thick. The branch section had fallen from a very tall Florida pine and it was lying conveniently close to the one-sided fight.

I scooped up the branch and gripped it tightly as I approached the bully from his blind side. He was so focused on beating Pete that he never saw me coming. When I was within striking range, I called out his name very loudly. Then, when he turned, I caught him across the upper midsection with my bat. He wasn't seriously hurt, but I had knocked the wind out of him, so he immediately crumpled and fell to the ground.

Sister Thaddeus, a stoutly built Dominican nun in full black and white habit had just walked onto the schoolyard to assume her duties as the playground monitor. Unfortunately, the only part of the fracas she witnessed was the part where I downed the bully.

Here, it is important to note that I loved Sister Thaddeus, who had been my 5th-grade teacher, and I know she cared about me. She must have believed that I had gone berserk because she rushed over, grabbed me by my shirt collar, and shouted, "Thomas Ross, what have you done?!"

Anytime someone uses the name that appears on your birth certificate, it is safe to assume that you are in trouble.

Sister Thaddeus spun me around and shouted, "I'm taking you to the principal's office!" And, with that, she whisked me away.

When I looked back, the bully was doubled up on the ground and was being surrounded by a growing gathering of students and another Dominican nun was wading through the crowd to help the downed student.

Shortly, Sister Thaddeus and I were in the hallway outside of the principal's office. She pointed to a "waiting chair" and told me to, "Sit right there! I will be back for you." Then, she disappeared into the office of Sister Mary Margaretta, the principal.

In Catholic school, going to the principal's office was a lot like a prison inmate being summoned to the warden's office. It wasn't likely that anything good was going to happen there.

After Sister Thaddeus had enough time to explain to Sister Margaretta what she had witnessed, she reappeared and said, "Sister Margaretta will see you now. You can go in."

Sister Margaretta was another one of the nuns of whom I was very fond. While she also wore a full habit with only her face and hands visible, she was immediately perceived as attractive. She was small, scarcely over five feet, probably smaller than Pete. Her features were delicate and her complexion smooth and fair. She could have easily served as a model for the Madonna. However, it would have been a mistake to use her appearance to judge her ability to serve as principal and disciplinarian of hundreds of young and, at times, rowdy Catholic kids. Sister Mary Margaretta had the demeanor of a six-foot two-inch tall, two hundred twenty-five pound—drill sergeant! And, you didn't want to cross her because you would lose any battle you started.

When I walked into Sister Margaretta's office, she had her head down and was writing on a sheet of paper.

"Probably writing a note to my Mom and Dad," I thought.

When I reached her desk, she looked up, pointed her pencil toward one of the two chairs in front of her desk, and said sternly, "Have a seat, Thomas. I will deal with you in a minute."

After a few minutes, which I feel certain were designed to let me stew, Sister put her pencil down, folded her hands, and looked me straight in the eyes. Then, the conversation went approximately as follows.

"Thomas, you have heard me speak about fighting on the

schoolyard. Correct?"

"Yes, Sister."

"How many times?"

Trying to minimize the event, I said, "Well, Sister, this wasn't so much a fight as it was a scuffle. It didn't last very long, hardly worth a visit to your office."

"Thomas . . . I will decide what merits a visit to my office!"

"Yes, Sister."

"I suppose you have a good reason for hitting Joseph with a tree branch?"

"I was just trying to stop him."

"From what exactly?"

I shrugged—then told her the truth, "Pushing kids around. Today, he punched my friend, Pete, in the face."

Sister Margaretta sighed; she knew Joseph was a bully. Then, "Thomas, I can appreciate you wanting to help your friends, but fighting for them isn't the answer. You do realize that, don't you?"

I knew exactly what she wanted me to say, but I wasn't quite sure how to answer in a way I truly felt without getting myself deeper in trouble. Then, it came to me. I asked, "Sister if someone was beating up on Sister Thaddeus, wouldn't you help her?"

The normally unflustered nun became silent, just looking into my eyes. I felt as if she were searching my soul. Then, "Thomas . . . Thomas . . . Thomas."

I knew she was giving thought to her response. Just then, there was a knock at the door. When she answered, "Yes? Come in," I couldn't have staged the next moment any better. It was Sister Thaddeus who opened the door! And, when she did, Sister Margaretta looked directly at me. I raised my eyebrows, shrugged again, and showed her a lopsided grin.

Sister just shook her head, got up from her desk, and went outside the office to see what Sister Thaddeus wanted to tell her.

A few minutes later, Sister Margaretta returned, sat behind her desk, and once again folded her hands. Then, "Thomas, Sister Thaddeus has been told by other students that she didn't see what they saw before you became involved in the "scuffle." She used my word, *scuffle*.

"Yes, Sister."

"Sister Thaddeus was told that if you hadn't intervened, Peter might have been seriously injured. And, you should know that I am going to call Joseph's parents and have them come in to visit. Joseph may be expelled for fighting."

Since it sounded as if I were going to be let off the hook, I said nothing more than, "Yes, Sister."

"Thomas, I want you to know that I still can't and don't condone fighting to solve problems. Do you understand?"

I stuck to my line, "Yes, Sister."

"Very well. Then you can go."

"I only varied a little, "Yes, Sister. Thank you."

Then, I got up and started out of the office. Just as I turned the door handle, Sister Margaretta called to me. "Oh . . . Thomas."

I was halfway out the door, but leaned back in, "Yes, Sister?"

"To answer your question . . . yes, I would help Sister Thaddeus if someone were beating up on her."

I shook my head knowingly, smiled, and said, "I knew you were a fighter."

Sister Margaretta just shook her head and dismissively waved me out of her office, "Go, Thomas! Go!"

Because this book was written for a very broad audience, I haven't been overly descriptive of the abuse described by the Montagnards. But, Don Tate, a war correspondent who was present during the rescue, would later write graphically about what he was told during an interview with one or more of the Montagnards. A few weeks after the rescue, and after his report had been released nationally, I was mailed a copy of Don's story as it appeared on Page One of the Pittsburg Press on August 12, 1968. His account described some of the horrific details shared with him. Sadly, barbaric things can happen during a war. With today's technology, I feel certain that the story can be located.

It was enough for me to know that women and children had been abused. Sergeant John Herbert, one of the American advisers assigned to My Loc when Mang Quang and the others first arrived, was told that some of the Montagnard women, mothers, allowed lice to flourish in their hair to keep the enemy soldiers away from them. Disgusting, I know. And, imagining my mother and sister in such a situation like the one described by the Montagnards made me angrier than I have ever been in my life. Having attended Catholic schools, I had heard more than once that vengeance fell within the providence of the Lord and that it was His to dispense. But I had never been exposed to such wretchedness and I was determined to end it—one way or another.

This time, when I went to help the Montagnards, I would have more than just a pine bat as an "equalizer." With me, would be some of the world's best trained military advisers and more than enough of Thieu ta's combat-experienced Vietnamese troops. And, we would be ferried into battle and protected by some of the bravest and most fierce aviators I have ever met. This time, I would have both American and South Vietnamese power and support as equalizers to make sure these bullies were dealt with appropriately. It was with these thoughts that I headed to the airstrip, prepared to lead our rescue team in a risky attempt to free the Montagnards. And, I feel certain that, if she had known about our mission, Sister Margaretta would have been cheering us on.

The American and South Vietnamese Flags

*These are the two flags under which other Americans
and I would serve our tours of duty.*

CHAPTER 7

The Enemy Wore Make-up

THESE THREE HALF-NAKED, half-starved mountain villagers have just given us a unique and welcomed opportunity, I thought. This wasn't going to be a mission designed to attack-and-kill or search-and-destroy. It was going to be a humanitarian effort, a rescue—a mission with an objective to save lives. And, any military man or woman will tell you that they would rather save a life than take one.

Continuing to consider what we were about to attempt, I realized that there was going to be something very personal in this mission for me. This was exactly the opportunity I had hoped for, but hadn't really expected and—had nearly forgotten. This would be my chance to do the "something good" I had come to do. I felt as if the Montagnards were offering me a gift.

How could we not help them? The courage and determination that had brought us together were admirable. While they had been beaten physically, they had not been defeated mentally. They were small men, but they were also husbands and fathers whose courage was as big as their love and concern for their families.

With a promise made, the true urgency of our mission immediately began to loom before me. And, it was obvious that Sergeant Koch realized the need for quick action when he asked, "How are we going to get everything done that we need to do and be out there by tomorrow, LT?"

I looked at him and said, "Well, Sarge . . . you and I are going to be very busy for the next few hours. And we need to get started right now."

If we were going to do something for the villagers and, in particular, Mang Quang's family, there was no time to waste. And, there was still much I needed to know. So, speaking to the interpreter but looking at Mang Quang, I said, "Tell him that before we can help his people, we need help from him. We need to know how many villagers there are and exactly where the village is located. We need to know how many soldiers are guarding the village . . . we need to know many things." As the interpreter translated my words, Mang nodded his understanding. With that, I took him outside where we had a commanding view of the surrounding landscape from the top of the outpost. This is where our mission planning would begin.

Trying to determine the location of the village turned out to be more difficult than I imagined. Mang Quang's response to my first question about where the village was to be found wasn't a great deal of help. After the question was asked, he walked to a large nearby rock, climbed up onto it, extended his arm, pointed out to the west, and said something I didn't understand.

When I looked at the interpreter, he said, "He says it's . . . out there."

As Sergeant Koch and I looked to the west, we were looking out over hundreds of square miles of varied terrain, some of it very rugged.

"Well, that's a start," I said optimistically.

Clearly, obtaining meaningful information from Mang Quang wasn't going to be easy. The interrogation would, however, turn out to be a near textbook scenario, just like the ones taught and demonstrated at the Intelligence School back at Fort Holabird, Maryland.

While serving as the Intelligence Officer of Company B, 3rd Special Forces Group at Fort Bragg, I took advantage of an opportunity to attend the Intelligence School at Fort Holabird, near Baltimore.

A "Top Secret" security clearance was required and the training I received at Holabird was some of the most interesting I experienced while in the military. There were about twenty young officers in the class when we reported for the first day of training. During the introduction to the course, we were shown things that looked like they were straight out of a James Bond movie. And, they were real. I had no idea such things actually existed.

Because we were all Special Forces officers who could be expected to work behind enemy lines, we were taught many methods of surreptitious entry—gaining entrance to places where you weren't supposed to be. Among many other things, we were taught how to defeat locks of many different types. After demonstrating techniques, we were all given a variety of different locking devices and they were all locked. As homework, we were to take them back to our quarters and bring them all back *unlocked* the next morning. There was one I thought about using a sledgehammer on, but I finally got it.

After being taught many other skills a Special Forces "operator" might require behind enemy lines, we moved on to the art of interrogation. And, no, none of that training involved waterboarding. Rather, it involved learning how to skillfully and artfully craft necessary questions to obtain needed information. In the case of Mang Quang, I need to ask basic questions that would make it easier for him to help me understand where our rescue team would find his village.

For several days, we were taught various interrogation techniques, then we reached the practical phase. The classroom was rectangular and had what appeared to be dark glass across the top half of the wall at the front of the room. We had all already guessed that we were being observed through two-way glass. And, we were half-right. We weren't being observed, but we were about to become observers.

When the instructor stepped to the podium, he explained that we would each have the opportunity to put what we had learned to practical use. He further explained that actors would be used as our subjects and that, depending on the scenario described, they would be dressed as local Vietnamese, VC, or NVA soldiers. Students were asked to assume that information in the script was real and to conduct themselves as though the actors were actually Vietnamese citizens or captured enemy soldiers with valuable intelligence information. The student's task was to gather as much information as possible from them using various interrogation methods we had been taught.

The interrogation would be conducted on the other side of the two-way glass set up to resemble a military field situation. The rest of the class was to observe the interrogation, then ask questions or comment during a debriefing and critique of the student interrogator.

When the first interrogation began, a description of the scenario was read to the student interrogator and the class. Then, the interrogator was taken into the interrogation area and put in position ready for his subject(s).

Then, when the first subject entered the room, he was an NVA soldier in full khaki uniform and he was escorted by two U.S. Army MPs in full field uniforms. I was immediately impressed with the realism. The enemy NVA soldier was actually an American Caucasian actor in make-up. But, even when I had my turn at interrogation and was up close, the actor's demeanor was convincing enough to make the experience feel very real.

During at least one of the interrogations, an actor posing as a VC soldier became very convincing and provocative, yelling foul names at the interrogator. He yelled things like, "You are American pig!" and, "You are baby killer!" We were all shocked when his student interrogator became so enraged that he went over the desk after the actor. He was ready to fight—then and there.

Those of us watching the interrogation burst out laughing when the actor fell backward, jumped up, and ran for his life. That's how realistic and intense the interrogation training could become. Ultimately, the class was extremely interesting and the value of its practical application was about to be proven.

5th Special Forces Group Flash

This 5th Special Forces flash was on the green beret of those who served in Vietnam.

When 5th Group returned to the United States after the war, the flash was changed. In the mid-80s, the yellow and red, which represented the South Vietnamese Flag, were removed and the flash was black with only a white border.

Then, in 2017, the flash was changed back to its Vietnam era colors. The reason for the change was to honor the 5th Group's history and, especially, to honor those who did not return. Having been a part of that history, I was very pleased with the change.

CHAPTER 8

Finding the Village

LESSONS LEARNED AT FORT HOLIBIRD would eventually help us find our way to Mang Quang's mountain village. But the process would be slow, long, and require patience—mine and Mang Quang's.

Mang Quang had given us a clue to the location of his village. He told us that it was "out there." To see if I could get a bit closer than that, I asked Sergeant Koch to get my map from the jeep and I spread it out on the hood.

Well, that turned out to be of little help. After showing the map to Mang Quang and trying to explain what it was and how it worked, he looked at me and shook his head. Not surprisingly, he had no idea how to read a map. Again, he gestured toward the west and repeated his original "out there" response, which was met with a chuckle from an amused Sergeant Koch.

Reliant on the Holabird interrogation training, I decided to begin a series of very basic questions. Some of them may seem silly, but the answers and Mang Quang's resolve to help would provide enough information to move the mission forward.

Looking west, one could easily see that much of the terrain was mountainous and appeared to be covered with thick jungle. And, I knew that to be the case because I had flown observation missions far to the west before just to know what was out there. The answer—nothing but jungle could be seen from the air.

So, anticipating the answer to my next question, I asked it anyway.

"If we flew like a bird 'out there' and we flew over the village, could we see it from above?"

I knew my guess was correct when Mang Quang responded by shaking his head negatively back and forth.

"He says, 'No. The jungle is very big and very thick. The village is very well hidden under the trees,'" the interpreter said.

This was meaningful information, so I obviously needed to continue with the more basic questions.

"Okay, good. Now, ask him to describe the land around the village. Is it flat? Are there hills?" Before I could say more, the interpreter understood what I needed to know and began the questioning.

"He says their village is in the hills near the base of some mountains, Trung Uy."

"Good!" I said smiling.

The smile was to let Mang Quang know that his information was helpful. It was certainly going to take a while to learn all the things we needed to know and I was concerned that he might become frustrated.

At this point, I won't attempt to list the many questions Sergeant

Koch and I asked Mang Quang. There were simply dozens and dozens of them, probably hundreds. So, I will merely demonstrate the questioning process used that day. It should already be apparent that the process was long and tedious.

After a few more questions, we felt confident that the village was roughly two days' walk away in mountains somewhere in the southeast, not near enough information to plan a mission. The necessarily long series of questions continued with Sergeant Koch taking notes and asking his own questions.

Every time Mang provided a piece of important information, I encouraged him, "Good, Mang Quang. Very good!"

Sergeant Koch added his encouragement by patting him on the back. Mang would always smile widely, obviously pleased that he was helping.

While he was indeed providing helpful information, the things he told us still narrowed the location of the village only to a vast area far to the southwest of Trung Dung. The exact distance he covered would depend on how many hills he walked over, how many and how far he had to walk around, how long he stopped to rest, along with a host of other key factors.

Pausing momentarily to consider my next question, I was half staring at my map. Then, my eyes focused above the edge of the map toward the direction of the Song Cai River, which ran through the valley. That gave me an idea. The village, to survive, would need a water source.

Maybe it's near the river, I thought.

Realizing the river could easily be seen from the outpost, I stood up from the customary squatting position we had assumed to talk. Turning to face the river, I motioned for Mang Quang and the interpreter to come beside me. Then, putting my hand on Mang's shoulder, I pointed toward the river and traced its flow through the valley until it disappeared in the west.

"Is there a river near your village?" I asked through the interpreter.

He smiled again and said, "Yes."

I picked up a stick and drew a long slowly curving line in the dirt.

"This is the river," I said.

Then, scratching an "X" on both sides of the river, I asked, "On which side of the river is your village?" pointing first at the one X and then the other.

"That side or . . . this side?"

Mang didn't hesitate. He quickly pointed to the X on the south side of the river. Then, he spoke as he continued to point.

"He says the village is on the hills in the jungle above the river, Trung Uy."

When asked if there was anything else around the village besides the river that would help us find it, we were given an important graphic response.

Mang Quang took the stick from my hand and scratched some marks of his own in the dirt. He drew some small inverted "V's" on the south side of the village. As he drew, he spoke to the interpreter and continued to draw some larger "V's" on the south side of the smaller ones.

"These are the small hills where the village is located," the interpreter said, pointing to the small Vs. Then, he pointed to the large Vs.

"He says these are the higher mountains beyond the village in this direction."

Then, without knowing it, Mang gave us the key piece of information we would need to locate the village from the air.

Still scratching with the stick, he drew several squares.

"What are those?" I asked.

"Cornfields, Trung Uy. He says there are cornfields around their village," the interpreter said as Mang handed the stick back to me.

Of course, I thought, *those must be the fields where they found the Chieu Hoi passes.* I thought for a moment or two studying Mang Quang's drawing before asking my next question.

"Are there any cornfields near the village other than these?"

"No," came Mang's very definite response.

Even though he was primitive with no formal education, it was obvious that he was very bright. Mang Quang seemed to know that he had finally given us enough information to find his village and his family.

Of course, there was still more to know, but at this point, Sergeant Koch and I finally agreed that we had enough information to find the village. As important as knowing its location, though, was the need to find a place to land close enough to the village to reach it quickly.

Sergeant Koch and I had both become concerned when Mang first described the village as being in the hills under the jungle canopy. We were afraid we would have to land some distance away and move through the jungle. That could take a great deal of time. If we were unable to reach the village quickly, the VC would have more than enough time to kill or move the villagers before we reached them. They would also have time to set up an ambush for us or, at least, prepare for our assault.

Now, however, we would have a choice of ready-made landing zones. The cornfields, which Mang Quang described as being very near the village, were an unexpected bonus and perhaps more than we could have hoped for. We would land in one or more of them and move immediately into the village and, hopefully, complete the rescue quickly.

With a reasonable idea of how we would locate the Montagnard village and how we would approach it, there was another extremely important issue to consider. Once more I turned to the interpreter. "Ask Mang Quang about the enemy soldiers. How many are there and where are they?"

Mang's responses were long and surprisingly detailed. Among other things, he indicated that an NVA unit had left the camp only a few days before his escape. He also told us that the absence of troops and some of the guards had made it easier for the trio to slip away undetected. Mang then said that, normally, the soldiers were posted on trails as well as in various places in the jungle all around the village. But, now, he believed there could be no more than thirty to forty men guarding the entire area. If that were true, it would be another unexpected break.

Sergeant Koch and I both asked several more questions regarding the disposition of the soldiers around the village. We also asked Mang and the other two men for their best description of the weapons carried by the enemy soldiers.

All Mang and his friends described were light weapons. It seemed the NVA unit had taken all the bigger pieces with them when they left the village. That was extremely welcome information. But then he shared something he had overheard just before being released that was alarming. Mang told us that he overheard one of the NVA soldiers telling one of the local VC soldiers that a large NVA unit would be passing through the village in approximately a week. That news suddenly made our need to act quickly more critical. It seemed we had a very narrow window in which to attempt a rescue.

Now, with all the other issues, we had to hope the NVA unit didn't arrive before we got there the next day. Still, the remaining unit of "thirty to forty" soldiers, that were known to be in the area, could prove very resistant. A well-commanded unit of that size, defending on its own jungle terrain, could inflict significant casualties on our force.

If a large portion of the enemy unit was concentrated in identifiable areas, an airstrike might help to eliminate the threat. But, according to Mang Quang, they weren't. And, because of the dispersal of villagers throughout the area, the possibility of using an airstrike was virtually eliminated for fear of killing one or more of them accidentally.

During the questioning, it had become clear that to be successful and hold casualties to a minimum, we would have to move swiftly and strike with surprise. And, we needed to do whatever we were going to do as early as possible the next morning if we were going to save Mang Quang's family.

Now agreed that we had all the information we needed to plan our rescue mission, Sergeant Koch and I also agreed that to be "out there" the next morning, I needed to get to work very quickly on the Operation Plan that would make it happen.

"Okay, Sarge. Let's head back to camp and I'll get started. There's still a lot to do and very little time to get it done." At that point, it was a little after 1400 hours (2:00 pm).

We loaded Mang Quang and the other two men into the jeep and started back to camp. We drove down off My Loc hill with considerable anticipation of what we were about to attempt.

Crossing Mountains, Jungles, and Valleys

This is the western view from the top of My Loc. Mang Quang and his two fellow villages had come from the far mountains that can be seen under the clouds in the distance. This was only a part of our view when Mang Quang told us that his village was "Out there."

Rugged and Remote Terrain

Much of the terrain in the area where I expected we would find the Montagnard village looked like this, rugged and isolated.

Map of "Out There" — "The Middle of Nowhere"

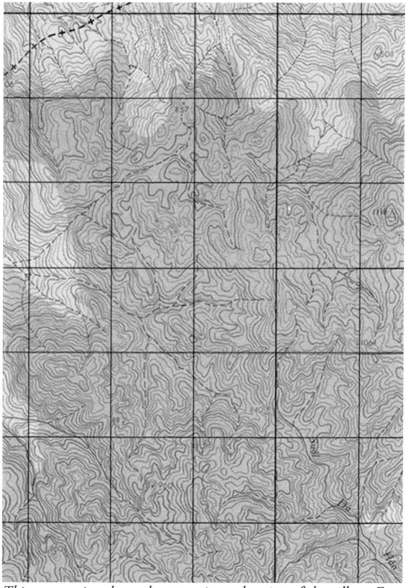

This map section shows the approximate location of the village. Even someone who can't read a map can see that there are no roads or signs. Almost my entire map looked like this. And, since agricultural features aren't shown on maps, the cornfields can't be seen—we would have to find them.

CHAPTER 9

Gathering the Team

TIME SEEMED TO BE PASSING quickly, and with so much to do, the speed at which Sergeant Koch was driving us back to Trung Dung suddenly seemed awfully slow.

"Is this as fast as we can go, Sarge? We really need to get back to camp."

"Are you saying you want me to go a little faster, sir?"

"Yeah, if you think this thing will go any faster."

"You've got it. Hang onto your beret."

Because the road was still wet from a morning rain, the wheels spun when Sergeant Koch changed gears and mashed the accelerator pedal. The Montagnards grabbed the backs of our seats to hang on as our speed increased and thrust us all backward.

"Sarge, you know this is a unique opportunity, but it's one full of serious challenges."

"Yes, sir. You're right. How are we gonna pull this off? It's too far to walk and make it on time. So, we're gonna have to use choppers. If the VC or NVA are there, even thirty or forty, as these guys say, we're gonna have to prep the LZ. Hell, LT, their people will scatter into the jungle along with the VC. They'll think we're shooting at them, too."

"Yep, you're right. There's a lot we're going to have to think about. That's why we need to get back to camp as quickly as possible."

Koch nodded, and we continued bouncing and sliding our way toward Trung Dung.

As we made our way back to camp, there were many things to consider. While enthusiastic about a rescue mission, I couldn't shake the thought that what we had been told might simply be part of an elaborate trap.

When we reached Trung Dung, my very first task would be to secure the troops required to be the major portion of my ground rescue unit. But, as American advisers, neither I nor any of my A-502 teammates commanded Thieu ta's Vietnamese troops, we simply advised them. So, to form a rescue team that I would lead, I would need Thieu ta's approval.

As soon as we reached the compound at Trung Dung I asked Sergeant Koch to find some help getting the three Montagnards cleaned up, clothed, and fed. Then, I quickly headed out to find Thieu ta.

Even though he was eating a late lunch in his quarters, he motioned for me to come in when he saw me at his door.

"*Ciao,* Trung Uy. Please . . . come in and sit with me. How are you today?"

"I am fine, Thieu Ta. How are you?"

"Good, good. I am good. You want to have lunch with me, Trung Uy?"

"No, Thieu Ta. Thank you. I won't be eating lunch today. But there is something I do need. I need some help."

"Tell me what you need, Trung Uy."

"I need about two companies of your best soldiers, Thieu Ta," I said.

Then, after I told him about the Montagnards and what I wanted to do for them, his only question was, "Is this mission important to you?"

When I assured him that it was and that it would give meaning to my service in his country, he paused and there was a brief silence as he seemed to give thought to my request. He then placed his chopsticks across his rice bowl, wiped his mouth with a cloth, and dropped it on the table. Then, he turned to face me.

"Trung Uy," he said, "you can have whatever you need."

"Thank you, Thieu Ta. Thank you very much."

"So, will you command this mission, Trung Uy?" he asked.

When I said, "Yes, I will," he surprised me by saying that he would come with me to command the Vietnamese troops, rather than simply assigning one of his Vietnamese officers as my counterpart. When I told him that he may want to reconsider because I had no idea what resistance we might face, he said, "If you're going . . . I'm going! We will go on this mission together since it is so important to you."

"Very good, Thieu Ta. I am glad you will be with us."

After thanking him again, I got up and turned to leave. Even with troops committed, there was still a great deal to do.

"Please excuse me. Now, I need to go find us a ride, but I will see you this evening to brief you and make final plans," I told him.

"Okay, good. Then, I will see you tonight, Trung Uy."

I saluted and moved on to the next piece of the puzzle.

Since his approval was required, everything hinged on Thieu Ta's sanction of the rescue mission and his commitment of troops. So, I was very pleased, but not at all surprised when he said I could have whatever I needed.

Ngoc and I had developed a warm relationship during my time at Trung Dung, beginning on Buddha Hill and, most recently, during our time together in the Dong Bo Mountains. Unexpectedly, the brotherhood we had formed would be rekindled years later, when even our families became close.

At the end of my tour in December 1968, I went to say good-bye to Thieu ta. I gave him a folded note with my parent's home address and phone number written on it and told him that they would always know how to reach me. Then, I invited him to call if he ever got to the United States. I never expected to see him again and hoped he would survive the continuing war.

After sharing a few memories of our experiences together, we shook hands and I left to catch my ride to Cam Ranh Bay and the flight home.

When the government of South Vietnam collapsed in 1975, Ngoc, his wife, and their four children fled the country with thousands of other refugees. They escaped with what they were wearing, almost no money, and not much more than a few phone numbers. One of them was mine.

After moving through various refugee camps, they were relocated to one in Huntsville, Alabama. Ngoc and his family reached Huntsville late in the evening. The next morning, he found a phone and placed a call.

I had just finished mowing the lawn at my Gulf Breeze, Florida home, and was walking into the house when I heard the phone ringing. When I answered and said, "Hello," the voice on the other end asked, "Trung uy?"

Even though it had been years since I left him at Trung Dung, I knew instantly that it was Ngoc and said, "Thieu ta?"

"Yes, I just wanted to let you know that I am with my family and we are in the United States."

While Ngoc's call was a surprise, I knew instantly that he must have kept the note I had given to him years earlier—and I was right. After speaking to my Mom a few minutes after Ngoc (she called me), I learned that he had first called my parent's home and my mother answered. She told me that before he finished explaining who he was she knew that it was "Knock, knock." She said that Ngoc seemed very excited and happy that she knew who he was. She gave him my phone number and told him that she and my father would look forward to meeting him. She didn't realize how soon that would be.

Although I had wondered about Ngoc and his family many times while watching the news of his country's fall, it had been nearly seven years since we last spoke.

Ngoc wasted no time telling me that the only way they could leave the refugee camp was if a sponsor could be found for him and his family. When he asked if I had any ideas, I told him not to worry. "You just found your sponsor," I said. Since Huntsville was about a six-hour drive from my home in Gulf Breeze, a suburb of Pensacola, I told him to expect me to arrive in about six or seven hours.

For a few weeks, Ngoc and his family stayed in my home.

During that time, we found him a job and a place for his family to live. Ngoc, as he wanted to be called, was able to make contact with some of his family living in France, and they sent him money to help him with his new start.

While they were living with me, Ngoc and his wife, Kim, who resembled a delicate china doll, woke up early every morning. They would clean the house and work in the yard. I often found them on their hands and knees picking weeds out of my flower beds, which was their way to return a perceived favor. Even though I made them quit cleaning or picking weeds whenever I found them doing it, the next morning they would be at it again.

When Ngoc left Vietnam, he was a colonel and held a very important position. He and Kim owned two homes and lived a respected upscale lifestyle, yet while they were in my home, they never hesitated to take the most menial tasks upon themselves. They became very special friends.

While living in the Pensacola area, our families also became close. Ngoc and Kim called my parents "Mommy" and "Daddy." Since then, we both moved several times and Ngoc and Kim's four children have had children.

As noted earlier, one of the last times Ngoc and I spent time together, he was—ironically—talking about returning to Vietnam. If the tables were turned and I had been forced to go to another country, I suppose I might be trying to make my way back to the United States. Thankfully, that's not something I have to be concerned about.

After leaving Thieu ta, I immediately headed for our radio room. I asked one of our radiomen to begin checking on the availability of air support while I went looking for a ride from the 281st AHC. We would need them to transport our rescue unit to the village. Their participation was essential if the mission was, literally, to get off the ground. However, the serious possibility existed that, for more than one reason, their involvement might be prevented.

On the way back from My Loc I was looking over my map and realized that, based on the information gathered from Mang Quang, it seemed apparent that the village would be found far beyond the western border of the map. Since that would be well outside A-502's AO, even though they were our assigned helicopter unit, the 281st would not be required to fly the mission. Additionally, because the Montagnards were civilians, the 281st might not technically be flying in support of an American military unit. That could be a problem if a military mission developed somewhere else and required the unit's choppers. Added to these issues were the facts that we weren't sure exactly where we were going, and we were uncertain about precisely what we would find when we arrived. So, a "We can't help you" response was a distinct possibility.

When the radioman answered at the 281st, I asked him if they were available to fly a mission early the next morning. He said there was nothing on the schedule but asked me to "Wait one" while he put his Duty Officer on the radio.

Shortly, another voice was on the radio and asked for the details regarding our mission request. I told him everything. I told him what we knew as well as what we didn't know. I also told him there would be absolutely no artillery support and, as of yet, there was no air support. No allied artillery would reach that far and I had just been advised that we were still waiting for air support commitment.

There was a long pause on the radio as the Duty Officer at the

281st was obviously considering his response. Then, "You know you're giving us pretty short notice for something this big, right?"

"I do. But this just fell into our laps, and we feel that we have to respond. And, if we are to have any advantage, we need to be there at first light."

There was no hesitation and there was hope in his next transmission. "I agree. In view of what you've told me, I think we have to be there early."

Obviously accepting the mission, he went on to assure me that the 281st was "all in" when he said, "Just tell me where you need us to be . . . and at what time. The 281st will be there."

Then, he continued saying, "I'll be flying one of the choppers myself. I want to be a part of this one, so I will be your Flight Leader."

I would later learn that the man on the other end of the radio at the 281st, was Captain John Wehr. And, his "You can count on us" type response was a typical one that A-502 and every other unit they supported came to expect from the 281st Intruders. They said they would be there—and they were there!

Turning to another puzzle piece, I began thinking about the problem of alerting the people in the village to our rescue attempt. Sergeant Koch had recognized a problem we would have to resolve.

Still in the radio room, I turned back to one of the radios, rotated the frequency knob, and called the Eighth Psy Ops headquarters in Nha Trang. When they responded, I asked if they had an aircraft capable of broadcasting a recorded message via loudspeakers. They told me the Air Force had one they used from time to time.

"Is it available for a mission tomorrow morning, and can you help record the message?" I asked.

The response was, "Yes" to both questions.

Putting A-502's various operations together was often like assembling a puzzle. Each one was a little different than the last or the next, but the process was always the same. You put the obvious parts together first. Then, the trick was trying to make all the remaining pieces fit. The next piece of this puzzle was one that would be key.

I hadn't worked with the Psy Ops people before, but I knew exactly how their plane would fit into the mission. It would be very important to the success of our operation.

Sergeant Koch had identified a significant problem regarding the operation on the way back from My Loc. His concern was valid about how we would alert the villagers to the rescue attempt. Because we would be firing on or around the landing zones to suppress any enemy activity, our action would likely cause significant alarm among the villagers.

Despite our worthy intentions, the villagers might be frightened by our aggressive arrival and flee deep into the jungle when they saw or heard us approaching—unless they were somehow informed of our intentions.

My solution to the problem identified by Sergeant Koch was to tape-record a message to the villagers that would let them know that our soldiers had come to rescue them, not to kill or hurt them. At first, I considered having our Montagnard interpreter tape the message. Then, I was struck by the logic of having Mang Quang do it. He could identify himself to the villagers and, hopefully, that would give credibility to the message as well as helping to calm the villagers as the sky overhead filled with helicopters. Mang Quang could also give specific and familiar directions to the pickup point. So, alerting the Montagnards to exactly what was happening would be the responsibility and the mission of the Psy Ops pilot. This particular part of the puzzle would fit well.

With parts and pieces falling into place, I collected Sergeant Koch and Mang Quang, and we left for Nha Trang. The drive was going to take a while so, with Sergeant Koch again at the wheel, I began to work on a script for Mang Quang's message and the timing of its use.

By the time we reached the 8th Psy Ops Headquarters, they were ready for us and we were ready for them. The message was prepared along with a plan for its broadcast during the mission. I had also finished a basic outline for the overall mission. Before making the tape, we needed to have a basic sequence of events outlined so Mang Quang could tell the villagers what was happening.

After we arrived at Psy Ops Headquarters, I asked our interpreter to translate the script and write it down for Mang Quang. Then, I reviewed my plan with the Psy Ops team and invited their input.

Our attempt to rescue Mang Quang's family and the other villagers would begin when we lifted off from Trung Dung's small airstrip at zero six hundred (6:00 am) the next morning. We knew generally where the village would be found, so we planned to fly west through the valley after liftoff.

We estimated the village to be twenty to thirty kilometers west of our My Loc outpost on the south side of one of the Song Cai River's forks. Since the villagers had not crossed a wide river on their way to us, we would simply stay south of any wide ones. Mang Quang's job would then be to pinpoint the location of the village once we found the cornfields, which would be our most important landmarks. Mang Quang was very confident that if we could get him to the cornfields, he or one of the other villagers could quickly lead us to the location of the village.

During our earlier conversations, Mang had described one particular cornfield that was on a hill above the village and higher in the hills than any of the others. Tentatively, I chose that field to serve as our primary landing zone, but of course, I would need to see it before committing our troops. If it looked good, it might also give us the advantage of holding the high ground if we encountered trouble. Mang Quang said that hill and field would also be an easily recognizable gathering point for any people who may not be in the village when our troops landed.

As we continued to review the plan, I explained that after we found the cornfield, we would mark it with smoke for the gunships. They would then make a couple of passes firing their guns and rockets to discourage the VC from interfering with our landing. If the gunships received no significant return fire, Mang's message would begin to be broadcast, and the slicks, lightly armed helicopters, would move in immediately to begin offloading troops.

As troops continued to be offloaded, Mang's message would tell the villagers why we were there and urge them to gather on the cornfield as quickly as possible. Once we had collected everyone, we would fly them back to Trung Dung—to safety and freedom.

Finally, before we made the tape, our plan was carefully explained to Mang Quang so that he would understand why and exactly what he was to do.

With the explanation completed, I asked Mang Quang if he understood what we were going to do and what he needed to do. He shook his head up and down, indicating that he did.

With Mang Quang and the Psy Ops team prepared to tape, we were ready to begin creating the message we would use the next morning. Eager to begin, Mang reached for the microphone.

Just before we began taping, and as the interpreter handed the microphone to Mang Quang, I asked him to tell Mang to use his own words when speaking his message. We needed to make sure that the villagers not only heard the message but would understand what they needed to do.

Mang and the interpreter rehearsed the message several times. After about the fourth or fifth rehearsal, Mang nodded his head and told the interpreter he was ready.

Then, we were ready to go "live." The technician who was making the tape for us turned on the recorder and motioned to Mang to begin speaking. Then, with no further coaching or prompting, he immediately began speaking. Mang was a natural.

It was interesting watching this primitive jungle dweller speak into a device he had never seen before. He seemed focused and spoke with what I took to be authority, as if he understood the importance and urgency of his message. As Mang Quang spoke, the interpreter looked at me, smiled, and nodded approvingly. Mang was doing exactly what he had been asked and coached to do.

The message I had written for him was brief:

> This is Mang Quang. This is
> Mang Quang. I have returned with
> help. The guns shoot at the VC, not
> at you. Move quickly to the high
> cornfield where the soldiers are
> landing. They will protect you.
> Move quickly!

The message was necessarily brief since the plane with the speakers was a propeller-type and would be flying fairly fast as it passed low over the thick jungle. The plane would have to fly low to ensure that the message could be heard clearly through the dense jungle canopy.

Additionally, even though the message was short, it wouldn't be possible for the entire message to be heard on one pass. The pilot would have to fly over the same area many times for the message to be heard in its entirety. If the terrain were as rugged as we expected, this mission would be very challenging and dangerous for the Air Force pilot.

When Mang Quang finished speaking, the interpreter seemed excited. "Perfect! He did it perfect the first time, Trung Uy." Even the interpreter was impressed by the primitive jungle dweller's effort.

"Good job," I said, patting him on the back and showing him a wide smile, the international symbol of approval.

He seemed pleased with himself, but I believe he was also

amused by what we had asked him to do because he began to laugh.

He had previously seen and heard the VC radios, but had never seen a tape recorder before and was uncertain about its use. He knew he was to speak into the small box we had provided for him and say the words he had practiced. But he hadn't fully understood why until we played the tape back for him a little later.

When he heard his voice coming from the recorder his eyes opened wide. He pointed at the recorder and, as if questioning, he repeated "Mang Quang? Mang Quang?"

He laughed a little, then put his hand on the recorder and began shaking his head and repeating his name softly.

When the tape finished playing, I had the interpreter explain that we would play his message loud enough for the villagers to hear when we went for them the next morning.

Mang Quang began shaking his head up and down.

"He says, this is good, Trung Uy," the interpreter said, "He says it is good his people will hear him."

By the time we finished the tape, it was early evening and we needed to get back to camp before dark. I still needed to coordinate and finalize plans with Thieu ta.

I thanked the Psy Ops technician who had made the tape for us. He said he was glad to help and that he would re-record the message in a loop so that it would play continuously while the pilot was flying over the village. Thanking him again, I handed him the information the pilot would need to link up with us the next morning. At the same time, he gave me the pilot's call sign.

"He's a very good man. You'll be glad he's with you."

"I look forward to meeting him," I said.

After exchanging salutes, Sergeant Koch, Mang Quang, and I got back in the jeep for the return trip to Trung Dung.

Occasionally, on the way back to camp, Mang would repeat his name and shake his head. Apparently, he was still trying to determine how his voice had been trapped in the small box.

When we reached Trung Dung, I asked Sergeant Koch to reunite Mang Quang with his friends and arrange for them to have something else to eat and a place to sleep. Then, I walked over to Thieu ta's to finalize our plans. When he asked if I had eaten yet, I told him that I had eaten breakfast but hadn't had time for lunch. He then insisted that I have dinner with him while we discussed the mission. Glad to accept his invitation, we spent most of the evening finalizing our plans.

After being confident we had resolved multiple aspects of the operation, including contingencies for the unexpected, I left Thieu ta and walked back across the compound to our team house to review the operation with the other U.S. team members who would be involved.

While it was in no way unexpected, the enthusiasm of our team members for what we would be attempting the next morning was remarkable; everyone was willing to go. However, the only ones going were the regular advisers to the units Thieu ta had selected for the mission. This meant that only five Americans, including me, would be going with our initial rescue team. Everyone else would remain at Trung Dung on "Standby" in case of an emergency requiring reinforcements.

By the time the team briefing was finished, there was little left of the night. It had been a very, very long day. With my map and notes in hand, I walked down the hall to my room to get some sleep. But, at the entrance to my room, I tossed the map and notes onto my bunk. Then, went out the side door of the team house. My walk took me out into the empty center quadrangle where I stopped and looked down toward the airstrip where our rescue attempt would begin in a short-few-hours. Our rescue team had been gathered and—we were ready to go.

The air was still as I stood, a single figure in the darkness, staring into the vastness of the night. There were no sounds except for the occasional and familiar bark of a dog in the nearby village. Looking up toward the stars, the only companions to this night's black sky, a few quiet moments were spent reflecting on this day and wondering what the next one would bring.

CHAPTER 10

Going "Out there"

SLEEP CAME QUICKLY AFTER falling into bed, but it didn't last long. The new day seemed to burst into being. Fading stars were joined by loud churning helicopters, the 281st was arriving—just as promised.

A thin orange line of daylight traced the eastern horizon and confirmed that morning was also arriving. I stood outfitted for the mission about where I had been the night before, just before going to bed. I watched the arriving choppers as they passed low over my head, one by one. Bright camp lights illuminated their olive-green underbellies and the spinning blades that held them in the early morning sky.

It was going to be a beautiful morning. The thin line of daylight was quickly growing into a shimmering yellow-orange glow over the eastern wall of the fort. The sky was clear except for a few small white clouds off in the distance. Our mission would have a good start. The weather was perfect—it was time to go.

Thieu ta's troops were already at the airstrip and our American advisers were conducting pre-mission checks. I motioned to Mang Quang and his two fellow villagers that it was time to go. They jumped down off the front of the jeep where they were perched like a flock of small birds, and we started walking toward the airstrip together. We hadn't walked far when Thieu ta and his driver came around the corner of one of the buildings in his jeep.

"Oh, Trung uy. Good morning."

"Good morning, Thieu ta."

"You want a ride to the airstrip?"

"Sure. Thank you."

I motioned Mang and the other two to get into the back of the jeep and jumped in beside them.

When we reached the edge of the airstrip, we stopped on a small rise that provided a slightly elevated view of the runway and got out of the jeep.

In front of us and slightly below us, ten troop-carrying helicopters were lined up with their blades spinning. Three gunship helicopters, which would fly protection, approached and landed at the end of the line. More than two hundred troops with full combat gear and an assortment of weapons were assembled all along the runway.

The collection of men and equipment at the airstrip created an image of power and strength. My senses became keen as I watched troops making adjustments to their equipment and considered what we were about to do. Significant amounts of adrenaline began pumping into my youthful bloodstream.

Scanning the runway, there was astonishing detail to everything that I saw and heard. A myriad of things was simultaneously perceptible. Our U.S. advisers worked with their Vietnamese counterparts to prepare the troops. Rifle bolts slapped shut and radio checks were audible as final equipment checks took place. Mixed with things that could be seen and heard were the alternating smells of helicopter exhaust and fresh air. The helicopter exhaust fumes would intrude on the fresh morning air blown in from the South China Sea. Then it would fade, only to return with the next wind shift. Much was happening, but nothing seemed confused. I don't believe I've ever felt more focused before that moment, maybe because I knew that my life and the lives of others would depend on how events unfolded that day.

As I stood still watching the mission preparation, I had a unique and amazing experience that has been very difficult to explain until now.

My memories of that unique moment and the experience remain clear and vivid, even though it has been more than fifty years since it occurred. I had felt the grip of fear and knew what it was like to believe my death could be imminent, but this experience was very different. While standing there on the small rise surveying the assembled force, I momentarily experienced a remarkably strange feeling.

Soldiers preparing for combat have experienced and described a multitude of feelings just before going into battle. Those feelings have ranged from near numbness to sheer terror, each easily understood. However, as I've said, what I experienced that morning has been more challenging to explain.

As my gaze rose above the troops and helicopters to the distant mountains, I felt someone's arm on my shoulder. For a moment, I thought Thieu ta had put his arm on my shoulder as if to encourage me. However, when I turned to see who had touched me, there was no one within three feet of me. Then, refocusing on the rescue team, I suddenly felt strangely different. In the first edition, the sensation was described as feeling—invincible! But that wasn't quite it. When I wrote this chapter many years ago, *invincible* was the best word that came to mind to describe how I was feeling at that moment.

Throughout the years, when discussing the feeling or asked about it, I've tried to find a better description. Then, one day a former teammate sent me an email that contained a brief block of prose, a poem of sorts. When I finished reading the words, my spine tingled as I relived the moment on the small rise above the gathered troops. What I had just read described my feeling that morning almost perfectly. And, as it turned out, it took more than one word to describe it.

The few lines sent to me by my teammate were written by an unknown author and I found them uncomplicated, but powerful in meaning. They involved fate, seemingly an entity, taunting a warrior. The poem or quote follows:

BEWARE THE STORM

FATE WHISPERS TO THE WARRIOR,
"YOU CANNOT WITHSTAND THE STORM."

AND THE WARRIOR WHISPERS BACK,
"I AM THE STORM."

THAT DAY, I was a warrior and I felt as if—I and our team were "the storm."

Despite all of the trappings of war before me, I felt absolutely certain neither harm nor death could penetrate whatever force had enveloped me. The feeling was exhilarating for the few seconds it lasted, and I marveled at its almost intoxicating effect.

Was the unique feeling caused simply by an excess of adrenaline being pumped into my bloodstream? Or, just a general feeling of invulnerability experienced at one time or another by nearly all youth—who knows? During my days as an altar boy, I could have easily been convinced that the Archangel Michael had settled in beside me with his sword in hand and wanted me to know he was there.

Imagination and romance aside, I believe that the feeling was more likely based on a belief that our cause was just and noble. After all, who could be hurt or killed on such a worthy mission of mercy?

Whatever the cause of the amazing feeling, I had the good sense to recognize my mortality and realize that believing you were invincible was a very good way to get yourself killed. Still, for the very short time it lasted, I savored the feeling—that would never be experienced again.

Regaining my grip on reality, I echoed Mang Quang's first directions to where we would find the village and the families, "Let's go, Thieu ta. They're . . . out there."

We walked the rest of the way down to the runway where Thieu Ta gave some directions to his unit commanders, then we went to the waiting helicopters to meet the pilots.

"Good morning! How is everyone?" I asked upon approaching the gathered aviators.

"Good" and "Fine" came the quick and crisp responses.

"Great! I'm Lieutenant Ross and this is Major Ngoc, our Vietnamese Camp Commander. We are very glad you could all be a part of this operation. Thank you for coming on such short notice."

Thieu ta quickly added his own welcome, "Yes, thank you for coming. Lieutenant Ross has created a big challenge for all of us."

"That's what we're here for," said one of the pilots as he stepped forward to introduce himself.

"I'm Captain John Wehr. I'll be your mission flight leader. My callsign will be Bandit Leader. I will also coordinate the Wolf Pack gunships for you." He then turned to another Captain standing near him and introduced Captain Ted Dolloff, the Wolf Pack Platoon Leader.

Captain Wehr seemed assured and confident, exactly what our mission needed. Later, it would become obvious why he was the leader.

"Excellent. It's good to meet you. Your assist with this is going to make the entire mission possible."

"Well, Lieutenant, there's something I'd like both of you to know. When our pilots and their crews heard this was a rescue mission, everyone wanted to be a part of it. I hope nothing happens anywhere else today. Our Rat Pack platoon is on standby in case you need more reinforcements than the Bandits can insert on our own. So, you have more than enough lift support available today. Just tell us what you need us to do for you."

"We appreciate your commitment. Here's what we think is going to happen," I said.

Then, spreading my map out on the hood of a nearby jeep, I began to brief the pilots.

The briefing started with an outline and explanation that once we located and marked an LZ, Thieu ta Ngoc and I would initially coordinate everything from directly overhead in Captain Wehr's "Bandit Leader" chopper.

While I knew Mang Quang wanted to be on the ground looking for his family and I wanted to be there with him, we needed to establish and maintain as much control as possible. My opinion was that, of the three Montagnards, Mang Quang was the strongest and knew the most about the area where we were going. Consequently, I wanted him to be with me and Thieu ta. One of the other two Montagnards would go in with the point unit while the third would remain at Trung Dung and serve as the guide for the reinforcements if they were needed. That man would also be available to meet the incoming villagers I expected and hoped to be sending back to Trung Dung.

The pilots were told that, because we were uncertain about the exact location of the village, flexibility would be the key to our success.

Captain Wehr immediately commented, "Well, in that case, you've got the right guys for that challenge."

There was some laughter from the other pilots as I continued. I explained that Bandit Leader's command chopper would remain mobile over the village after arriving in case something unexpected occurred and plans needed to be changed quickly. After inserting our troops, the other choppers would fly back to Trung Dung and be ready to return with reinforcements. The gunships would also remain over the village to provide air cover.

Hopefully, since there was so much unknown to our mission, this operational plan would allow me maximum flexibility in command and control. We would also simultaneously survey the other cornfields as alternate LZs for incoming troops or reinforcements. If reinforcements were required, Thieu ta and I would land and lead them into wherever we were needed.

Point by point, I went through each step of the operation as it was expected to unfold, down to our return to camp. In finishing the briefing, I mentioned the most important part of getting the mission started.

"The key to getting on the ground and finding the village," I said, "will be locating the cornfields. So, as we head west, whoever sights them first, please alert Captain Wehr in his command chopper. When we locate the cornfields, our villager will point out the one that is closest to the village. If it appears viable as an LZ, I will ask Captain Wehr to have the Wolf Pack light it up (fire rockets and miniguns) before the insertion of our troops."

I paused, then asked, "Are there any questions?"

When no one responded, Captain Wehr repeated, "Any questions, guys?"

Everyone shook their heads, indicating there were none.

"I guess not," he said. "We're ready to go."

"Good, so are we. We just need to wait for . . ."

My remark was interrupted by the high-pitched sound of the engines announcing the arrival of the Psy Ops speaker plane as its pilot made a low pass over the field.

I finished what I had started to say, ". . . that guy."

As he pulled up over the end of the runway, he checked in on the radio.

"Bunkhouse Zero Two, Zero Two. This is Tracer, this is Tracer. Over."

"Roger, Tracer. This is Zero Two. Are you ready? Over."

"Roger. I'm loaded and ready to talk. Over."

"Good. Are you clear on the mission, Tracer? Over."

"Roger. Understand I'm to stay with you to the target and follow your direction for the delivery of my payload. Over."

"Roger. Tracer, that's it. Stand by. We're coming up to meet you."

Turning back to the pilots, "That's everyone. We're set. Let's go find 'em and bring 'em back!"

The briefing broke up, the pilots started back to their choppers, and I asked Thieu ta, "Are you ready?"

Testing my resolve, he asked, "Sure you want to do this, Trung uy?"

I laughed and said, "Yes . . . I do. It's too late to back out now."

He smiled and said, "Okay, then . . . let's go."

I folded my map, which I expected to be useless, and stuck it in my web gear, and said, "Okay, then. Thieu ta, you can load our troops."

Getting Ready to Go "Out There"

Above foreground: A-502 advisors are ready to begin the mission. Background: With my flight helmet under my arm, Captain Wehr and I are making last-minute plans to head into enemy territory in search of the village.

The Last of the Bandits Arriving

Captain John Wehr, Bandit Platoon Leader, would direct the 281st's air units and his helicopter would severe as our command ship.

O-2 Cessna "Skymaster" — Callsign, "Tracer"

Major Ken Moses, the pilot of this Air Force "Psy Ops" aircraft, would prove to be as valuable as any member of the rescue team about to be assembled. He flew high in the sky until he was called upon to deliver his payload. Then, he dropped out of the sky and skimmed the jungle canopy just above the treetops.

The Wolf Pack — The Eye of Our Storm

The early morning arrival of the three Wolf Pack gunships, one landing and two in trial.

Wolf Pack's Minigun and Rocket Pod

This is one half of the Wolf Pack gunship's weapons system. Captain Ted Dolloff would direct the Wolf Pack and Captain Bain Black would operate the rocket and minigun weapons system.

Some of Thieu ta's Troops — Ready to Go

Vietnamese troops were eager to help me get the job done. Thieu ta told them this mission was important to me and they should do their very best—and they did!

Time to Go!

Thieu Ta has just given the order to load. It was time to head west.

While Thieu ta passed the order down the line for his men to board the choppers, I had Mang Quang and the interpreter get aboard Wehr's chopper. Ngoc and I stood on the runway beside the lead helicopter until the troops who were going with us finished loading.

We weren't taking all the troops gathered at the airstrip. A backup force would remain behind and ready on the edge of the runway in case we had been lured into a VC trap or received unexpected resistance. They would serve as reinforcements who could be moved in quickly. This rapid-response team was one of my contingencies for the unexpected, hopefully, one we would not be required to use.

Once the choppers were loaded, Thieu ta and I climbed aboard and I plugged the cable on my headset into the communications outlet on the chopper. This, as usual, would permit me to communicate with the pilot and other U.S. units involved in the operation. Thieu ta had his radio, which would provide him with communications to his Vietnamese units. A quick radio check with everyone was accomplished while the choppers powered up. Then, it was time.

"Zero Two, this is Bandit Leader. We're ready."

"Roger, Leader. Let's go."

With its blades cutting through the hot, humid Vietnamese morning air to create lift, our helicopter began to rise off the ground. It moved forward down the runway picking up speed. Then, dropping its nose slightly, we rose effortlessly up over the outer edge of the runway and the trees beyond.

We flew over the village outside the camp, made a slow turn to the west, and headed out through the valley.

I motioned for Mang Quang to move over by the door with me so he could see out and help direct our flight. With the wind whipping his jet-black hair and clothing, he didn't seem to be comfortable that close to the door. But, with a death grip on a piece of the chopper's framing, he nodded affirmatively when asked if he was all right.

On our way west, we would fly directly over My Loc. The outpost would serve as our first point of reference for Mang Quang on the reverse trip to his village. Then, with him acting as our guide and using the information he had given us the day before, we would attempt to retrace the route he had taken to reach us.

We flew along the south side of the Song Cai River on our way to MY Loc and passed the Korean artillery base, which was on the north side of the river. A portion of the White Horse Division, the unit occupied the top of a hill that had been bulldozed so that there was little or no vegetation on the hilltop, which was smooth and rounded. Five or six rows of barbed wire ringed the base to provide some protection against a ground attack. Their artillery was inside the wire and pointed skyward in various positions around the hill.

Unfortunately, Colonel Chang's Korean guns would do us no good this day, though. The artillery we had so often been able to count on in our normal area of operation wouldn't be able to reach us this time. As mentioned earlier, we were going so far west that we would be far beyond the range of any "friendly" artillery.

The only immediate fire support would be the three Wolf Pack gunships that flew above and behind us at a slightly higher altitude. But knowing their reputation, I felt very comfortable with them as our guns.

Korean Artillery Base

Members of Trung Dung and the Korean base became friends because we visited the other's base, shared intelligence information, and supported each other on various operations.

Other air support had finally been committed. Spooky in Nha Trang and a pair of Air Force F-4's at Cam Rahn Bay had signed onto our mission and were available on a "Standby" basis.

Spooky was a modified C-47 that belched a hailstorm of blazing munitions from sets of Gatling-style miniguns. The F-4's could drop a range of bombs from napalm to high explosives.

The only problem was that air support of any kind was typically on a first-come, first-served basis in Vietnam. Even though Spooky and the F-4s had been alerted and were prepared to support our operation, if another unit or operation got into trouble, they would properly be dispatched to assist that unit or operation.

That standby air support situation only made the operation a bit more challenging. It also made an important statement about the American advisers who had volunteered for this mission, knowing that it might have to be completed with the individual skill and equipment they each carried. The amazing thing was that every adviser at Trung Dung wanted to be a part of the mission. Even my battle-worn buddy Phalen wanted to stick his neck out yet again and asked, "Why can't I go, too?"

"You've been into enough stuff." I said, "Just stay here and clean my room if you can't find something else to do."

As things would unfold and before the mission ended, every member of A-502 would have the opportunity to become a part of it.

Before we flew over My Loc, I asked the interpreter to remind Mang Quang about his role. I would show him the outpost where he had first found our soldiers. Then his job would be to try to show me the way back to his village.

After the interpreter finished speaking, Mang responded by eagerly shaking his head that he understood.

As we neared My Loc, I looked down on the rice paddies surrounding the small outpost and thought about Dale Reich, the lieutenant who had been killed during a night ambush. The site where his unit had been set up was along one of the paddies below. It came to mind that the enemy unit responsible for his death might be the same one responsible for holding the Montagnards. That unit had approached our ambush from the west, the same direction from which Mang Quang had traveled in his search for help. If it was the same enemy unit, they had been heavily armed the night of the ambush. During our questioning of Mang Quang, he indicated that they were away from the village on some unknown mission. Hopefully, that would still be true.

We passed almost directly over, but a little north of, My Loc so Mang would have a wider view of the valley. After pointing down to the outpost, it was a little surprising to see him lean out the door to look ahead. He pulled his head in and looked back at My Loc, which was behind us by then. Again, he leaned out the door to look forward. Then, slowly looking back toward My Loc once more, he moved his arm back and forth along our direction of flight, seemingly indicating that we were going the right way.

I looked out both sides of the helicopter for the other choppers. Five slicks were flying in echelon formation trailing back to the right and four back to the left. We were at the point of the "V" formation. Still trailing and slightly above us, there were two gunships a little farther out on the left and one the same distance out on the right. Tracer's speaker plane was above everyone and a little behind us. Our mini-air armada was an impressive sight. Occasionally, I could see it reflected in the water of the rice paddies below.

As we flew west out through the valley, Mang continued to lean out the door. Occasionally, he smiled as he pointed toward things on the ground and told the interpreter he knew where we were. I was glad, because there were a lot of people, his and ours, depending on his ability to find the way.

More than once, I hoped that Mang's recognition of terrain features was accurate and that he was guiding us correctly. Since he had never flown before, things must have looked very different to him from the air, but he seemed quite confident as he guided us over the jungle below. Occasionally, he would look ahead, down, back, and then gesture with his arm for slight changes in our direction of flight.

Farther ahead in the valley, hills began to rise from what, behind us, had been flat valley floor. The large valley we had followed began to split and forked into some smaller ones. I moved up to the opening between Captain Wehr and his copilot.

"See anything yet?"

"No," Wehr said, "but we're about to come up on twenty klicks past My Loc. Isn't that where you thought our search pattern should begin?"

"Right," I responded.

"Okay. I'm gonna pass the word to the rest of the flight."

"Roger, go ahead."

He pressed his radio switch and alerted the other pilots.

"This is Bandit Leader to Bandit Flight. We are entering our target area. Let me know if you see anything that resembles a cornfield."

There was a tug on my shirtsleeve. When I turned around, Mang Quang was pulling me back to the door. He began pointing toward a valley that doglegged off to the left of our direction of flight. He seemed excited. When I saw a river in the valley he was pointing toward, I got excited too.

"Leader, our guide wants to go up this valley that doglegs off to the southwest."

"Roger, I'll bring us around."

We had barely entered the smaller valley when Tracer called. Because his plane was faster, he was ahead of us and considerably higher than we were. He could see farther out in front of us.

"Zero Two, this is Tracer. Over."

"Roger, Tracer. Go."

"I think I've got your cornfields. There are several patches on the low hills out at about eleven o'clock. Over."

"Roger, Tracer . . . we're looking."

I moved back up between the pilots.

"Did you hear that?" I asked.

"Roger."

Captain Wehr, who was seated in the left front seat, saw them first. "Ah . . . yeah, I've got 'em."

He put two fingers up and pointed out through the Plexiglas window. Then I could see them.

"Yes, I see them, too."

Then I pointed them out to Mang Quang, who very enthusiastically indicated that we were indeed on target.

"Leader, this is Zero Two. Our guide says that's the place."

"Okay. We'll move in a little closer and then I'll orbit the slicks until we mark the one you want to use as our LZ. Is that okay?"

"Roger, that's fine," I said.

As we flew into the valley and toward the cornfields, I let Tracer know that he'd found our target, "Good work, Tracer. That appears to be the place."

While Wehr gave instructions to the other chopper pilots, I told Thieu ta we had found the cornfields and that he could alert his units. Then, while he called his men, I alerted the other four U.S. advisers who were each on four of the other nine choppers.

When we were close to our target, Captain Wehr gave his instructions for the slicks to break off, circle, and hold their positions.

"This is Bandit Leader. Everyone on my left, break, and orbit left. Everyone to my right, break and orbit right. Wolf Pack, stay with me."

After the choppers broke away and formed two spinning circles behind us, Captain Dolloff and his gunships moved in closer to provide protection as we began to descend toward the widely spread-out cornfields.

I called Tracer to keep him with us.

"Tracer, Zero Two. Stay with us but maintain your altitude above us. Take this opportunity to look the ground over."

"Tracer rogers."

I moved back over near Mang Quang, whose map in the sand had been as good as any printed on paper. Looking out the door toward the cornfields, it was obvious immediately which one was the high one. Mang and I pointed to it at the same time. We looked at each other and I gave him a thumbs-up.

I don't think he understood the gesture because he smiled, grabbed my thumb, and began shaking it.

"Okay, Leader, we've got the LZ. It's the highest one on that ridge at about ten o'clock."

"Roger, I see it," he said.

"Let's check it out."

"Okay, we'll take you in low. If it looks good to you, throw your smoke."

"Roger. I'll be ready."

Out of the side doors, the gunships could easily be seen. They were staying right with us. It was like having three big brothers with you to deal with any bully problems.

"Leader, this is Zero Two. When we mark the LZ, if we don't take ground fire, I'd like you to go ahead and have the Wolf Pack prep it for us. I want to insert our team as quickly as possible. If we do take fire and can identify its source, that will become Wolf Pack's immediate target. Over."

"Roger, understand. I will advise them."

As we drew nearer to the cornfield, Wehr prepared for our approach.

"Wolf Pack Leader, this is Bandit Leader. We'll make this a wide turn. Line up behind me and we'll approach a staggered file. When the LZ is marked, I will advise and it's all yours. Over."

"Roger, Bandit Leader. Understand. We're lining up on your tail now."

We completed the turn and started down with the three gunships that were in a staggered line behind us. Our crew chief, Jay Hays, and door gunner moved their .30-caliber machine guns back and forth over the jungle that was rapidly rising up beneath us. They searched for any threatening targets as we closed the distance to the cornfield. We were quickly getting low enough to the ground to see leaves on the jungle trees, which had been individually undefined from a higher altitude. At the same time, we were well within the range of enemy ground fire.

Removing one of the smoke grenades from my web gear, I moved Mang Quang and positioned myself back at the left door behind Wehr's seat. Looking ahead, the cornfield was clearly in sight. Through the jungle mist, it appeared steeper than we had hoped, but it would serve our purposes. As we made our approach, I pulled the pin on the grenade and took a very tight grip on the top of the door. And, as I did, I heard Thieu ta yell, "Don't fall out!" He wasn't kidding, he was serious.

I leaned out far enough to ensure the grenade wouldn't hit anything on the chopper and prepared to throw it.

We had kept our promise to the small Asian man at my side, and now we were "out there."

Wehr pushed his chopper into a final approach, diving directly toward the cornfield. It was early in the morning and, as clouds of fine white mist were still rising from the steaming jungle beneath us—we struck!

As we swooped low across the center of the cornfield, I threw the smoke grenade. "It's gone!" I shouted and Wehr immediately pulled up and away.

We quickly gained altitude as we circled back to the north. Looking out the door across from me, red smoke could be seen billowing up from the cornfield. The Wolf Pack gunships were given the all-clear to attack. By pouring fire onto what would soon become our LZ, I hoped a dramatic show of force would intimidate any nearby VC or NVA, allowing us to land our troops safely.

"Wolf Pack Leader, this is Bandit Leader. Your target is marked. Over."

"Roger, we're going hot now."

During the pilot briefing at Trung Dung, I explained to the pilots, and especially Captain Dolloff and his Wolf Pack Pilots, that I had no idea where the villagers would be when we arrived. In case they were near the cornfield chosen as our LZ, I asked that they do their best to keep their fire within the perimeter of the cornfield and not into the surrounding jungle. Sounds crazy I know, but I don't want to kill any of them during an effort to save them. But I do want the prep of our LZ to be an intimidating show of force.

Given clearance to attack, Captain Dolloff dove toward the cornfield. Flying with him was Captain Bain Black who also served as both pilot and gunner. Black would soon take over as the Wolf Pack Platoon Leader, so he was flying with Dolloff to learn as much as he could.

Captain Black immediately began firing into the mist and smoke. Rockets traced their paths with trails of white smoke until they disappeared into the swelling cloud of red smoke where they exploded upon impact with the cornfield. Pieces of cornstalks, dead tree stumps, rocks, dirt, and dust were blasted into the air, creating swirling brown clouds of debris.

Black's miniguns roared with a deep hum and the impacting rounds raised clouds of brown dust.

As we circled behind the gunships, Dolloff and Black were pulling up and out after his attack. I asked Wehr to let him know that, if they took no ground fire, we needed the gunships to clear quickly. I wanted to get troops on the ground and secure the LZ before the enemy could regroup.

"Wolf Pack Leader, Bandit leader. If you guys don't take ground fire, one pass each will be enough. Over."

"Roger, understand. We took none. If the next two ships aren't hit, we'll move out of the way."

The wolves howled as the next two gunships attacked one behind the other, firing both miniguns and rockets into different parts of the cornfield. Even over the whop, whop, whop noise of our chopper, ID number 113, we could hear the loud hum of their miniguns and the explosions of their rockets. The Wolf Pack was certainly giving me the show of force that I requested.

I couldn't imagine what the Montagnards must be thinking. But it would be easy for them to believe that hell was certainly being unleashed from above. I hoped the enemy felt that way and would clear out.

At that point, it was almost time for Tracer.

"Tracer, Zero Two. Can you see what has happened down here? Over."

"Roger, my seat is front row."

"Good, when that last gun is clear, we're going to make a pass so that our guide can point out the village location. When we have that location fixed, we will direct you to it, over."

"Roger. Tracer is standing by."

When the last gunship emerged from the huge cloud of dust and smoke that hung over the cornfield, we headed back down so that Mang Quang could point out the village position.

As we again dove toward the cornfield and jungle surrounding it, Mang Quang, who had earlier been uncomfortable with flying, was now leaning out the door with his face in the wind like a dog out a car window. He was smiling broadly and looked triumphant as he pointed an outstretched arm down into the jungle. Leaning into the wind, he appeared bold as he led us to his village.

After completing our flyover of the jungle, we regained altitude, but there was a problem. We would have to make one more pass before we were sure the problem was resolved because, on the first one, Mang pointed toward three places. They were separated by hundreds of meters.

At first, I thought Mang Quang was disoriented or confused, but when he pointed to exactly the same three places on the next pass, I began to suspect that our mission had just been made more complicated.

Even though the answer was anticipated, I told the interpreter to ask Mang why he was pointing to different places.

As it turned out, he wasn't confused at all. What Mang referred to as his village was actually divided into three separate living areas on two distinct terrain features. They were in the same general location, but one portion of the village was on the same ridge, but below the cornfield, and two were on the ridge just a few hundred meters to the west of the cornfield. Mang had identified all three for us. This was an unexpected, but important detail.

While we gained altitude after our last pass, Thieu Ta and I quickly adapted our plan to the new circumstances. Since we weren't sure what was waiting for our team, we decided not to divide our troops and send them in three widely separated directions. And, until I knew how much resistance we would encounter and where it would develop, I didn't want to commit the reinforcements waiting at Trung Dung.

We decided to have our team go first to where Mang Quang indicated that the largest portion of the village was located. That was the one slightly below and west of the cornfield. Then, depending on what happened there, we would either send them to the other two or call in the reinforcements, land in one of the other cornfields, and attack from a different direction.

Regardless of what happened next, at least we knew we were in the right place and, hopefully—at the right time.

US Army Air Medal

During the Vietnam War, the US Army awarded the Air Medal to Warrant Officer or Commissioned pilots and enlisted aircrew based on actual flight time. The award was also presented to infantry troops who flew on combat assault missions and those who met other requirements.

While I had no interest in collecting medals as a result of my military service, I was genuinely excited the day the Air Medal (pictured above) was delivered to me via U.S. Mail at my home in Pensacola.

The medal arrived with an accompanying citation explaining that it was being presented for missions "over enemy territory" totaling more than 25 hours. It amused me that an infantry officer was being awarded an Air Medal. Time on the Montagnard mission put me over the total hours required for the award—one of many reasons to remember it.

The reason I was excited was because this was my favorite of the medals awarded to my father for his service with the 8th Air Force during World War II. I thought it was bold and handsome. And, I loved the attacking eagle. But, most of all, the medal represented a special link between my father's service and mine.

CHAPTER 11

Go Find Them

WITH OUR ORIGINAL PLAN adjusted a bit, it was time to bring Tracer down, "Tracer, Zero Two. Over."

"Roger, this is Tracer. Go ahead."

"Tracer, it seems the village is spread out. One part is about two hundred meters below and west of the cornfield and two others are on the ridge west of the field. One of those is at almost the same level as the cornfield, and the other is roughly two to three hundred meters farther down the ridge. Over."

There was a pause as Tracer scanned the ground.

"Ahh . . . Roger, I have the area marked."

"Good. When you get down there, make several passes over those locations. Then, when you see the slicks approaching, work a little farther out until the troops are on the ground. Then, come back in. Over."

"Roger, I'll make room."

"Okay, Tracer, the stage is yours. We're calling the slicks in now. Over."

"Roger, Tracer is coming down. My speakers are live . . . now."

Then, like a hawk after a mouse, Tracer dove from his perch

high above us. Mang Quang's voice could be heard booming from Tracer's speakers as he flew past us on his way down.

With Tracer and our troops all headed for the cornfield, we were beginning to ring the bell of freedom, I hoped the Montagnards would hear it.

Hearing my conversation with Tracer, Captain Wehr radioed the slicks and pulled them out of their circling formations. He gave them directions that would put them on a course to the LZ. The cornfield was now officially our LZ.

Moving back to a seat in the center of the chopper next to Thieu ta, I told him the slicks were on the way.

"Good," he said, "my men are ready."

Looking out the door, we could see Tracer beginning his run. He passed very low across the ridges just below the field. Then he pulled up and made a looping turn and passed over them again. When he made his next run, I couldn't believe it when he disappeared down over the hill on which the cornfield was located. He flew out of sight down between the ridges. When he emerged on the other side of the ridge, it appeared as though he was coming almost straight up from the river bottom.

Tracer continued to make pass after pass over and between ridges, down small valleys, and through the river bottom. I was sure he was going to rip the wings off his plane on one of his passes through the jungle. He wasn't flying over it as he flew through the river bottom. He was down in it, below treetop level.

Later, when I asked him why he flew so low, he said he wanted to make sure the villagers could hear the tape.

Tracer's demonstration of flying acuity was no less impressive than that of the Air Force pilots who attacked the Dong Bo hilltop. He wasn't up in the wild blue yonder where the Air Force anthem suggested he should be. Instead, he was as low and close to the villagers as he could be without putting his wheels down and landing.

With Tracer's passes carrying him farther away from the field as the slicks approached, Captain Wehr took us up to a higher altitude. From there, we would be out of the way and Thieu ta and I would have a better vantage point from which we could see everything as it happened.

As they neared touchdown on the LZ, I became concerned for the first few slicks. If we had been lured into an ambush or if the VC and NVA were simply holding their fire until troops began to offload, the ones going in first would be the most vulnerable.

1st Lieutenant Mike Sullivan, who had been responsible for different areas at A-502, was on the first slick into the LZ and would be the first U.S. adviser out on the ground. He was the perfect man to have down there. Mike was a serious, experienced soldier, having been in-country for almost a year. He would provide stability for a tenuous situation and was responsible for organizing security around the LZ as troops on other incoming choppers continued to be off-loaded.

Sullivan was from Columbus, Ohio, and was a stocky, barrel-chested young guy who was an intense and very focused person. Even while he was in Vietnam, he lifted weights to stay in shape. He seemed to love being a member of Special Forces. Like most of those I met in Vietnam, he was a dedicated man, and I was glad he had accepted the challenge of advising the point unit, which would likely be the first to encounter any enemy resistance. I was certain that Sullivan would prove to be a fierce opponent for anyone unfortunate enough to challenge his unit.

From a vantage point several hundred feet above the humming activity below, we could see the slicks making their final approach to the LZ. They had white stripes painted on top of their blades that made them stand out against the steaming, dark-green jungle. They looked like spinning tops as they moved over the jungle canopy.

The cloud of smoke that first covered the LZ had cleared when our troops began hitting the ground. Circling just above the landing slicks, we were low enough to easily see everything that was happening. Immediately after the first 281st slick landed, we could see Sullivan and his unit fanning out across the cornfield, now our LZ. I scanned the edges of the open area for flashes, which would indicate enemy gunfire. But there were none—at least, no automatic fire. A couple of the slicks reported hearing semi-automatic gunfire, but they were unable to determine its location.

The lack of serious enemy fire was welcomed, but not altogether surprising. If, indeed, there was only a small enemy unit down there, disclosing their position could have meant immediate death. That was because the three Wolf Pack gunships were alternately swooping in, providing very close cover for each landing slick. Rockets or heavy gunfire would have rained in on any VC or NVA troop bold enough to make his location known. They would surely have been on the receiving end of the "Hell from Above" their motto promised and they had already demonstrated.

While the gunships provided protection for the off-loading slicks, Tracer, who was unarmed, continued his daredevil passes over the treetops and through the valleys farther away from the LZ.

"Tracer, this is Zero Two. You're doing a great job. Don't hit a tree!"

"No problem, I'm trying to figure out how to make money doing this."

When I told him to "Join the circus," we could hear him laugh as he once again disappeared behind a ridgeline.

Focusing my attention back on the LZ, the last slick was coming up out of the cornfield. It turned east and followed the others already headed back to Trung Dung, where they would land and wait on standby with the reinforcements in case help was needed.

"Bunkhouse Zero Two, this is Zero Eight." It was Sullivan. "We've got the LZ secured and all units are on the ground. We're beginning to move now. Over."

"Roger, understand. Good luck and keep us posted."

"Roger. Zero Eight, out."

This was a critical time. Would the villagers come to the cornfield or flee in fear? Temporarily, at least, the villagers would be in a position of terrible conflict. They were already on the ground with people who had abused them and who they knew to be their enemy. Now they were being confronted by unknown soldiers who were descending upon them from the air. I hoped that they could hear and understand Mang Quang's familiar voice being broadcast from Tracer's speakers and would try to reach the cornfield as his message directed.

With the slicks gone, Tracer moved back over the village and Captain Wehr acted as Air Controller for all the crisscrossing aircraft—us, Tracer, and the three Wolf Pack gunships. While he watched the air, I studied the lay of the land, paying particular attention to each of the other cornfields. Actually, being able to see them was far better than guessing about them.

Sullivan, with his Vietnamese counterpart at his side and one of the three villagers acting as guide, directed the rescue units to begin moving down the ridge toward the village. Once the units moved off the cornfield and disappeared into the dense jungle, they were very much on their own. With the slicks now back at Trung Dung, an immediate extraction or reinforcement, if necessary, would take some time. And, the thick jungle into which the units had vanished obstructed the vision of the pilots in the Wolf Pack gunships. If an enemy unit was encountered, the fight would likely be intense and at very close quarters.

We could do little for the men on the ground after they moved into the jungle, except pray for their safety.

Thieu ta and I circled above the gunships that were moving their very low passes farther down the ridge. They worked like a team of hungry wolves searching for prey as they attempted to intimidate any VC or NVA who might be in front of our advancing units.

Tracer's passes were on the ridges above, below, and on each side of Wolf Pack's swirling movement. While they had never worked together before, the movement between Tracer and the Wolf Pack seemed choreographed.

The ground units hadn't been out of sight long when Thieu ta's radio came alive with chatter. I was still looking down on the jungle when he tapped me on the shoulder to get my attention. He held up a finger for me to wait. I was sure he must have been receiving information from his men on the ground.

"Trung uy, they have some people," he said. Then, pointing to Mang Quang, he repeated, "My troops have found some of his people."

At about the same time, Sullivan called with a similar report. "Zero Two, this is Zero Eight. They're coming to us."

Amazingly, within approximately twenty minutes of the first troops landing on the LZ, the villagers were hurrying to meet them. Mang Quang's message had obviously been heard. Tracer had delivered it through the dense canopy to the jungle floor.

Later, after the operation had been completed, one of the American advisers told me that when one small group of villagers approached his unit, a couple of them were pointing into the air and saying, "Mang Quang, Mang Quang." It seemed clear that they had heard the tape being broadcast clearly enough to recognize Mang's voice.

The news of Mang Quang's people meeting our search teams was surprising and unexpected news, and I wanted to share it with him. When I called his name, he turned from the door where he had been anxiously watching the activity beneath us. I asked the interpreter to tell him that his people were coming to the field. Mang seemed to question the interpreter, then he looked at me and smiled and shook his head to confirm that he understood. After hearing the news, he quickly turned back to the door. His gaze seemed locked on the cornfield.

As Mang looked for the villagers not yet emerged from the jungle, the interpreter told me that he had asked if his family had been found.

If the villagers were indeed coming to us, Thieu ta and I decided it would be better to move our units back to the LZ. In case the villagers were coming from all three village locations as well as work sites in the area, it made more sense to consolidate on the LZ and provide a secure base where they could be protected. Since our rescue attempt had begun far better than expected, creating a strong collecting point would be preferred to walking around in the jungle looking for the villagers. If they didn't come or weren't able to reach us, we would then move units back out in whichever direction seemed appropriate and resume our search. Additionally, with the LZ strongly secured and covered by Wolf Pack, if our rescue unit were attacked, the entire team would be available to defend or counterattack.

"Zero Eight, Zero Two, over."

"Roger, this is Eight. Go."

"Eight, your counterpart is now receiving orders to move back up the hill. Make your base on the LZ. We're going to have you wait there since the villagers seem willing to come to you. Over."

"Roger, understand. We're turning around now. Out."

Sullivan and the rescue units moved back up to the LZ to wait for new arrivals. And, sure enough, in they came. In twos, threes, and small groups of fours, fives, and sixes, the villagers made their way up the hill to the LZ.

We couldn't believe what was happening. Despite the tactically poor layout of the village, the mission was running very smoothly. When we saw villagers emerge from the jungle and begin to gather on the LZ, the gunships were backed out a short distance so we could go down for a closer look. We flew down and circled the field at treetop level and, as we did, Mang Quang became very excited and waved to the villagers below. To let them know where he was, he began yelling his name and continued to wave. I'm not sure they could hear him, but some seemed to recognize him. They began pointing and waving back toward our chopper.

The sight of villagers waving at Mang Quang was remarkable. It made me feel good and very glad we had been given this unique and exceptional opportunity. I just hoped things continued as smoothly as they had begun.

With Mang still waving, we began to regain altitude so the Wolf Pack gunships could return to their close cover of the LZ. They were still behaving very much like big brothers and immediately swooped in as we cleared the area. When we pulled out, I asked Captain Wehr to take me to the other cornfields so I could see exactly how they looked in case we needed to use one or more of them. I numbered them and made notes.

After we had been over the village for some time, Captain Wehr began to be concerned about the gunship's fuel supply since they had done "a lot of fuel-burning flying." After checking with Wolf Pack Leader to confirm their fuel status, he suggested that they return to be refueled one at a time, which would leave two of them with us all the time. Wolf Pack Leader acknowledged and began sending the guns back one by one to have their tanks refilled.

We stayed on-site, watching as villagers continued to emerge from the jungle and gather in the cornfield. While we circled overhead, I had Sullivan send out small patrols to sweep the area around the cornfield. I wanted them to make sure the VC weren't moving in and told them to look for villagers while they were out.

We remained overhead until all the gunships had returned fully fueled. Then, we made the trip to have our own tanks refilled.

Upon our arrival back over the cornfield, Sullivan advised me that the flow of villagers had stopped. He also said patrols were no longer encountering villagers. When I asked him to take a count of those who had already arrived, he reported back shortly that eighty-two were with him on the LZ.

After Mike's count, the interpreter was asked to advise Mang Quang of the number and ask him if he thought that was everyone.

Mang said yes, he thought so, but he wasn't sure. He explained through the interpreter that the villagers had been kept separated for a long time. Some had escaped, and others had been killed or taken away. Since he had been closely guarded and his access to the other villagers had been restricted, he wasn't really sure how many were still in the area.

The assembled crowd on the ground appeared to be large enough to fill a small village. So, with no solid number to estimate the total number of villagers in the area, and since no others were arriving or being sighted, I asked Wehr to alert his Bandit slicks that we were ready to begin the pickup.

When the slicks began to arrive, we watched as the lead chopper made its approach to pick up the first group of villagers. As it neared the LZ, I called Tracer to release him for return to base and thank him for a truly incredible job, but he didn't want to leave.

"I'd like to stick around until they're out, Zero Two."

"Roger, stay as long as you like. We're very glad to have you here."

Then, when the first slick began to touch down, Wehr urged all of his Bandit ships to make quick pick-ups and departures.

"Bandit Flight, this is Bandit Leader. Get in there, get 'em, and get out. We don't want to lose them now. Over."

The approaching Bandit chopper, responded immediately, "Roger, Leader. They won't spend another night out here. Over."

"Roger, they're yours to take care of now."

As each of the remaining slicks began to touch down, the guns swooped in and out to provide the same very close cover they had given when the troops were offloaded. This time the slicks were carrying out a precious cargo of men, women, and children, but they were more than that. They were families who now had a chance of life in freedom without fear for their safety or their lives. With Mang Quang's family among them, it was clear by watching what was happening on the LZ that the men of the 281st weren't going to let anything happen to any of them.

We were all pleased with what had occurred, but Mang Quang seemed the happiest of all. His face almost glowed with the knowledge that his family was safe.

Captain Wehr became a little concerned that the gunships were staying too close together and too close to the ground.

"Bandit Leader to Wolf Pack Leader. You guys are pretty tight down there, aren't you?"

"Roger, Bandit. We are. Just covering our families in case Charlie decides he'd like to keep them here. Over."

Wehr laughed and responded, "Roger, Bandit understands. Take care of your families."

Listening to the exchange between the pilots, I realized how personal this mission had become for all of the Americans involved.

Tracer didn't want to leave until the villagers were fully extracted, and the Wolf Pack was going to hover over them until the last one was out.

During my tour of duty, I experienced a number of unique feelings. Some, I hope, I will never experience again. While certain emotions are more difficult to describe than others, the feelings I experienced while listening to the exchanges between these very seasoned combat pilots are fairly easy to describe: just as with my teammates, I felt incredible pride to be serving with men like these.

Every man involved in the operation knew what his individual mission required. From the time the mission began it was clear that a serious battle would be required to keep them from accomplishing their given tasks. They had come to find families and take them to safety—precisely what they had done. It had been a great day to be an American soldier.

Some of those Flown to Freedom

On the first day of the rescue, more helicopters were flying in and out of Trung Dung than at any time before. All of the activity created curiosity in and around the fort. Dien Khanh locals gathered on the fort walls to watch what was happening.

Mothers with Babies and Small Children

For the first time in years, women and children were free of abuse. At first, some seemed uncertain about how to act. But after a couple of days, they relaxed and began to smile and laugh. And, they seemed to enjoy all the attention they were receiving.

Thieu ta helped me arrange for food and housing for the villagers while our medics tended to medical needs and other teammates attended to clothing and other basics.

Yanked into the 20th Century

While free, the villagers were abruptly introduced to the modern world. They left the jungle in helicopters that flew higher and faster than birds, I couldn't imagine what they were thinking.

After teammate, Al Morace (above center), helped deliver food, this young girl clutched tightly to two bags of rice that she said were for her family. Some of the children looked very healthy, while others were boney and appeared starved. Our medics immediately examined those.

CHAPTER 12

No Joy

WITH ALL THE VILLAGERS back at Trung Dung, the last of our troops were picked up. Our small air armada turned east and headed home. Everyone was feeling very good about what had happened. We were able to find the cornfield with relative ease, the troops had been inserted quickly and without incident, the villagers responded to the broadcast far better than we had hoped, and everyone had been extracted without a single casualty. Villagers would tell us later that enemy soldiers had hidden in the jungle, fearing they would be killed. The rescue had been an extraordinary success.

Considering the events of the day as we churned our way back to camp, I was amazed and thankful that our rescue effort had gone so smoothly. When the choppers dropped us off at Trung Dung, however, we discovered that the results weren't as good as we first thought.

As soon as we cleared the spinning helicopter blades, Mang Quang, with me close behind him, went directly to the area being set up for the villagers. He couldn't wait to see his family. I followed him through the crowd as he searched for their faces. He was so excited, he looked as though he might burst with joy before he found them. But, while others celebrated their newfound freedom, this dedicated and caring husband and father would experience— no joy.

After making several passes around and through the crowd of villagers, Mang wasn't able to find his family. They weren't there. For whatever reason, they hadn't reached the cornfield and were still out there.

I asked our U.S. adviser who was keeping the headcount if all eighty-two were in the assembly area. He said he had just counted and that they were all there. He was sure because our medics were bringing gear down to do a medical check on everyone, and they had asked him to confirm the number.

Mang Quang wilted like a dying leaf before my eyes when he was faced with the agonizing reality that his family was still missing. He fell to his knees, grabbed his head with his hands, and started saying things I couldn't understand. Then, he began to cry.

His heartache was compelling. I quickly turned to the interpreter.

"We're going back! We're going back! Tell him quickly that we will go back for his family. We will return with the morning sun."

When Mang's crying changed to sobbing, it became difficult for the interpreter to understand and translate what he was saying.

"Trung Uy, he says his family will die before morning. The VC will know he brought the soldiers, and they will kill his family. He says you could give him ten million piasters (Vietnamese currency), but he would give it back. Without his family, it would mean nothing to him."

At that point, I felt as if my own heart had been ripped out. The exhilaration of the day had instantly turned to frustration and deep disappointment. While eighty-two freed Montagnards stood all around us, we had clearly failed this man who had prompted our actions.

Watching and listening to Mang Quang there on the ground in front of me, it was obvious that he believed his family would not live through the night. He wanted no wealth or much else without his family; even life seemed of little importance to him at that moment.

While the urge to return to the village right then was strong for all of us, I couldn't consider that possibility. It was already too late in the day. By the time we readied everyone to relaunch the mission, it would be dark when we reached the cornfield, assuming we could find it in the dark. Even if we could find it, we didn't need to be searching through the unfamiliar jungle at night. Furthermore, we didn't know the terrain as the enemy did, so even a small VC unit could inflict heavy casualties on us. Returning to the village before dawn simply didn't make sense.

Kneeling beside Mang Quang, I put my hand on his shoulder. "Mang . . . Mang Quang."

Slowly, as he swiped at his face with his hands, he turned to look at me. His face was covered with tears and dirt rubbed off from his hands.

I spoke to him softly, but confidently.

"Your family will be fine." Again, my interpreter began translating immediately without being asked. I spoke slowly so that the interpreter could translate simultaneously as I attempted to console the small, crumpled man.

"They are probably hiding in the jungle. I am very sure our

planes and soldiers scared the VC. They had to protect themselves. That would have given your family time to run away and hide in the jungle." Then, pausing for a moment, I continued.

"Mang, we will not leave them out there. The VC will not expect us to come again so quickly. When the sun returns to your village, we will return with it. And we will find your family."

He seemed to calm, so I kept talking.

"Mang, do you see these men behind me?" I asked, gesturing toward Thieu ta and several other American team members who had gathered around us.

"They say with me that the VC will not keep your family. Tomorrow night, your family will be here and they will be with you, and they will be safe."

When the interpreter finished translating my last remark, Mang looked at me and, as if testing the degree of my conviction, "Yeah, Trung Uy?"

Those were the first words he had spoken to me that I understood.

"Yeah, Mang Quang," I said with feeling and a much stronger voice. There was no need for the interpreter to translate.

In my attempt to make Mang believe, I had convinced myself that we would find his family. But a few moments later, after considering things said in my zeal to console him, I became very concerned. I wasn't sure what would be said to him the next day if his family couldn't be found, had been hurt, or—the worst—killed. What words could be offered to explain why I had been wrong about their safety? All that was certain to me at that moment was the need to keep him busy until the next morning.

Once again addressing him, I said, "Mang, in the morning we will help you find your family. But, now, we need you to help us.

We need to take care of the villagers who are here. Will you help us do that?"

After translated, he nodded, took another swipe at his face with each of his forearms, and, together, we stood up. I explained to Mang, who was regaining his composure, that I needed to go make plans for the next day, but would return a little later.

One of our medics who had been on standby for casualties was standing nearby and had heard what I said. He immediately stepped up and said he would take care of Mang and would keep him occupied. The medic handed him some medical supplies and asked him if he would help with the care of the other villagers. Again, he nodded. Knowing he was in very good hands, I patted him on the back and started for the radio room.

Walking away from Mang Quang, I realized something very important I had just learned from him. A man's capacity to love and care about his family has very little, if anything, to do with how primitive he may act or appear. If anything, the importance of family may be even greater to someone without all the trappings and distractions of modern society.

I hadn't gone far when Thieu ta caught up with me. He wanted to make sure I had realized something else when he said, "You know returning to same place is very dangerous . . . right, Trung uy?"

"Yes, Thieu ta, I know. But there is no way I can just forget that his wife and children are still out there. If we don't go back, I will be haunted for the rest of my life."

"Okay, I understand. Then, we go back together," he said.

"No, Thieu ta. That isn't necessary . . . you went today. I know you have an important meeting at LLDB High Command tomorrow and someone needs to direct care for the Montagnards. I have arranged for shelter, food, and medical care, but I need you to help me find a permanent place for them to live . . . maybe with another tribe somewhere. I can't do that, but you can."

I knew Thieu ta's meeting with General Quang was one he couldn't miss. So, after some discussion, he said, "Take whatever you need tomorrow. I will come back from Nha Trang as soon as I can. I will stand by here with the reinforcements. If you call for them, I will lead them."

He then cautioned, "Be careful, Trung uy . . . I don't like this."

Heartbreaking Disappointment

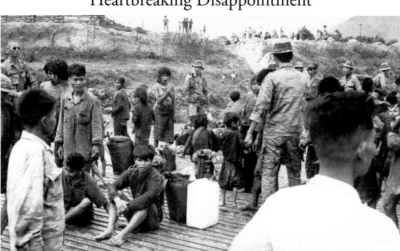

This picture brings back very sad memories for me. Mang Quang, Thieu ta, and I had just gotten off Captain Wehr's chopper. Mang was very excited about reuniting with his family. That's him in the bottom left corner of the picture as he begins looking through the group of Montagnard villagers. I was right behind him and in only seconds after this picture was taken, we would discover that his family wasn't there.

When he heard that choppers were landing at the airstrip with villagers, my friend, Lieutenant Bill Phalen, grabbed his camera and went to document whatever was happening. Thanks to him, I have these pictures. Unknowingly, with this picture, Bill documented both success and failure.

CHAPTER 13

We're Going Back!

WITH YET ANOTHER PROMISE made to return to the mountain village to find Mang Quang's family, there was much to do. After parting with Thieu ta, my first stop would be our radio room.

On my way to our small communications center, several team members who had heard we were back in camp stopped me along the way. They all wanted to know where the Montagnards were and what they could do to help. They were all told to see the docs (our medics), who would need a great deal of help.

Finally reaching the radio room, my first call went to the 281st. The Bandit flight had been back on the ground in Nha Trang only a short while at the time of my call. I asked for Captain Wehr and quickly told him the situation. After hearing my explanation, there was no need to ask for anything. Captain Wehr gave a direct and simple response.

"Tell me where you want us to be . . . and when."

"Same place, same time . . . tomorrow morning," I said.

"We'll be there."

It was that easy.

With the 281st signed on to another mission, Tracer was the next person who needed to be found. After tracking him down, his response was very similar to the one given by the 281st, "I know where," he said. "Just tell me when."

With Thieu ta's commitment of troops, the 281st, and Tracer again signed on to a second mission, our return to the Montagnard village had been quickly resolved. I left the radio room and started back to find Mang Quang and check on the Montagnards. I had just gone past the team house when Specialist Miller, the radioman on duty, called out to me.

"Lieutenant Ross."

"Yes. What's up? If your radio is broken, I don't work on radios."

Miller chuckled a little and said, "There's a Lieutenant Orians on the horn who says he needs to talk to you."

"Okay, let's go see what he needs."

Frank Orians was the PIO (Public Information Officer) for the 5th Group in Vietnam. He was a nice guy who I liked and had worked with from time to time. Whenever he had a VIP visitor at 5th Group Headquarters who wanted to see a Special Forces camp, he would often call me to arrange a visit and briefing.

"Roger, this is Bunkhouse. How can we be of service? Over."

"So, what have you guys got going out there? Nha Trang's buzzing about some kind of rescue 502's got underway. Is that true? Over."

My conversations with Frank were always a little less military in nature.

"Roger, it is. We brought out eighty-two Montagnards today. They tell us they've been used as slaves by the VC and NVA for several years."

"Damn, I wish we had known this sooner. We need press coverage on this," Frank said.

"Well, you've still got time. We're going back tomorrow. We've just discovered that there are still other families out there. If you have media available, they're welcome to come out and cover the story," I answered.

"I don't know who's around, but I'll find out and call you back."

"Frank, I have to tell you, because of what's being reported back in the States lately about search-and-destroy missions, this is the kind of story the American public really needs to be told. Everyone involved in this operation is putting his life on the line for these people. It's happening out in the middle of nowhere, in unfriendly country with little or no support."

"Okay, okay, it's late in the day. Let me get on it and see who's available. What time are you leaving tomorrow?"

"Zero six hundred. That's six in the morning for you."

"Yeah, yeah. Okay, I'm going to work on it right now. I'll get back to you before morning."

"Roger. If you find someone, you can send them out with the 281st. They're providing our ride."

"Good, good. I'll get back to you as soon as I can."

"Roger, I'll be right here. Zero Two, out."

This development was unexpected and fortuitous. Recently, there had been a great deal of international coverage about search-and-destroy missions carried out by U.S. troops in Vietnamese villages. One such mission in a village named My Lai would become infamous and synonymous with all of the worst occurrences in Vietnam. American soldiers mercilessly killed all or most of the inhabitants of My Lai.

Search-and-destroy missions required U.S. troops to go into areas suspected of being controlled by the VC or NVA to search for the villages from which they might be receiving support. If and when such villages were located, they were destroyed, usually burned to the ground. Unfortunately, to save the lives of loyal South Vietnamese as well as those of American soldiers, the seemingly obdurate action was deemed necessary. However, the indiscriminate killing of unarmed civilians was not policy and was absolutely unnecessary.

Because there were few good stories to tell, everyone at A-502 knew that media coverage with its often-graphic pictures of combat missions couldn't be doing much for morale back at home. Even we knew that negativism regarding the Vietnam War was mounting. News and mail from home made it obvious that since the Tet offensive of January, support for us and our efforts had begun to deteriorate more rapidly. We felt our story was a good one, and we hoped Frank Orians could find someone with the media who would tell it.

As the screen door to the team house slapped shut behind me on my way back to the Montagnards, the last thing on my mind was press coverage. If we got it, we could share a good story with family, friends, and our country. If not, nothing was lost since the deed was infinitely more important than the story.

When I reached the Montagnards' new temporary home, they were being fed dinner. I had the interpreter ask them if everything was good. When he did, many smiled and almost all shook their heads up and down while adding various guttural sounds of approval. This was probably one of the best meals they had enjoyed in years.

I asked our senior medic, Sergeant Jerry Arrants, if his team had checked everyone. He said that they had seen almost everyone and would check the few remaining before they left. I thanked him for his effort and walked over to where the Montagnards were being served. Taking a bowl of rice, I went with Mang Quang and sat cross-legged on the floor with several of the villagers. Mang seemed much more relaxed than when I left him. However, his thoughts were obviously on those far to our west, and he would not be at peace until his family had been found.

I stayed with the Montagnards until sometime after dark. Several times during the evening Mang Quang would go to the far end of the long shelter and look out into the darkness to the west. He was silhouetted against the darkness by the light of a nearby campfire as sparks from the fire flickered and crackled into the black night sky. It was fairly easy to guess what he was thinking about as he stared off into the distance. When I thought about that night years later, when I had my own family, I could only begin to imagine how he must have felt.

It was time to go back to the team house when the Montagnard mothers began putting their small ones down for the night. Preparations still needed to be made before the next morning when we would once again try to find Mang's wife and children and other families that we now knew remained somewhere in the jungle surrounding the village.

At the team house, I found my way to the refrigerator and helped myself to a cold Coca-Cola. Then, sitting at a table near the radio room, I began writing the outline of the operation for the next day's mission.

A good portion of the outline was finished along with most of my Coke when I heard Frank Orians's call come in over the radio.

"I'll get it," I said to Miller, pushing my chair back.

He gave me the handset as I walked through the radio room door.

"This is Bunkhouse. What's up in the city tonight? Over."

"Ha! Not much. That's for sure, and I know. I've been all over trying to find a news team. There's no one in town."

"Well, don't worry about it. You did what you could. We're going back with or without a news team and whether or not our effort ever receives coverage."

"No, I know it," he said, "but I hate it. You were right, even Zero Six [06 was the call sign for Colonel Arron, the CO of 5th Group] feels this is an important story that we want to tell. And we are covering it. We began working on a release this afternoon. So, I'll try to catch up with you late tomorrow to see what happened."

"Roger. I hope we have more good news for you. The family of the man who was our guide today is still out there, and we all really want to find them. That situation is truly a very sad one. Over."

"Well, I wish you luck and hope you find them. Good luck,

Zero Two. You and your team take care."

"Roger, no problem. I'll talk to you tomorrow. Zero Two, out."

Regrettably, it appeared there would be no media coverage. I hadn't even thought about the possibility before the mission, but teased with the prospect, I was disappointed. There had been so much negative reporting on the Vietnam War in 1968 that I felt some good news couldn't hurt. *This would be an excellent opportunity to share a little positive news with our families and our country,* I thought. Even my mother had asked in a letter if the war was as bad as it appeared on the news. She said that every evening the national news began with a count of how many Americans had been killed or wounded that day. She said, "I can't watch the beginning of the news knowing you are there."

If it had been covered, the story of the Montagnard rescue mission wouldn't have been a report about killing, wounded, or search-and-destroy. Instead, it would be a good story, an uncommon tale, and the kind that rarely reached the States. Nonetheless, since their presence had never been anticipated, the absence of the press would make little difference to anyone involved in the mission. A new commitment had been made to a small, dark-skinned Asian man by a small group of Americans who were going to see that it was met. We were going back—whether or not anyone else ever became aware that we made the trip.

The 281st's "Rat Pack" — Arriving from the South

All the clamoring troops gathering on the runway and the sound of thirteen arriving helicopters roused local villagers from their early morning sleep. It wasn't long before a large crowd had gathered all around the runway. Everyone seemed very interested in the military operation and nearly all stayed most of the day.

As the Bandits had the day before, the Rat Pack choppers returned to Trung Dung and waited on the runway after inserting our team on two cornfield LZs near the village. They remained there in case reinforcements were needed.

The Wolf Pack — Always There

CHAPTER 14

On the Rocks

THE FIRST LIGHT OF MORNING had just begun to brighten the eastern horizon. I was sitting on the hood of a jeep at the airstrip looking over my notes when elements of the 281st's lift platoons dropped in from the south. Dirt, grass, and dust blew everywhere as they settled down on the runway in a single file. Then, as they had the day before, Wolf Pack gunships cut circles in the humid morning air as they whirled in over Trung Dung. Jumping down off the jeep, I went to meet Captain Wehr so we could go over changes in what we had done the day before.

The blades were still spinning when I reached the lead chopper. I was surprised to see a different face.

"Good, morning. It's good to see you. Where's the other guy?" I asked.

"There was a last-minute change, and we'll be taking you out today. We're the Rat Pack!" the new flight leader said proudly of his unit. Then he introduced himself as Captain, Glen Suber, the Rat Pack's Platoon Leader, and said. "Our guys wanted in on this thing yesterday, so we're all glad to be able to get in on the action today."

"That's great. Welcome, I'm glad to have you be a part of this."

As the rest of the pilots walked up, I again welcomed the new team.

"Good morning. Welcome to A-502. We're glad more of you are going to get in on this."

There were several "Us, too!"–type responses, and then we quickly started the briefing.

My first comments were directed to Captain Suber, my new flight leader. "Let me go over some things that will happen today."

"Okay," he said.

"See that guy over there?"

"The major?"

"Yes. His name is Lee, and he's my boss. He'll ride out with you. Sullivan, the lieutenant who . . ."

Just then, Mike Sullivan walked up. "Oh, here he is right here. Good timing, Mike. I was just about to tell them you would lead the flight out."

"That's right," Mike replied.

Then, again speaking to Captain Suber, "While you're out there just make sure the boss gets a good look at what's going on."

"Okay, fine. We'll make sure he sees it all."

"Good. Then, as soon as you get Lieutenant Sullivan and his troops in on the LZ we used yesterday, send the slicks back for me and my guys."

"Okay."

"We will go in on another cornfield I saw yesterday. It's near a place where our Montagnard guide believes his family might have gone to hide. From there, we will sweep southeast through the portion of the village on the western ridge and link up with Lieutenant Sullivan on the original LZ."

Confirming his understanding of what was to occur, Suber recapped.

"Okay. So, we're taking them out and inserting them wherever you were yesterday. Then we're to come back for you, and you'll show us where you want to go in. Correct?"

"Exactly, and there are some other things you and the other pilots need to know."

Since Captain Suber and the Rat Pack were new to the mission, we discussed many things, and the pilots were given the opportunity to ask questions. We had just finished the briefing when Tracer roared overhead.

The radio on Ahat's back came to life with Tracer's check-in.

"Bunkhouse Zero Two, Zero Two, this is *T-r-a-c-e-r*, over." He drew his call sign out with a little more flair than he had the day before. He seemed to be feeling the fresh exhilaration we were all experiencing with the prospect of finding Mang Quang's missing family.

"Roger, Tracer. This is Zero Two. You sound ready to go again this morning. Over."

"Roger, get those things crankin', and let's go get the rest of 'em!"

Looking at the group assembled around me, I mimicked Tracer.

"Okay, let's go get the rest of 'em."

Then, I turned to Sullivan.

"Go ahead, Mike. They're all yours. Take 'em out."

"Okay, we'll see you out there."

Everyone climbed onto the choppers. Major Lee got on Captain Suber's chopper, and Mike Sullivan was on the second one back. Sullivan, with the Montagnard who had been with him the day before, would take the lead once airborne to direct the flight west to the cornfield. With the standard serious look on his face, Mike gave me a thumbs-up sign as they lifted off.

Feeling left behind, I watched as, one by one, the Rat Pack cleared the village on the south end of the runway and turned west. When the choppers were no longer visible in the distance, I went to Mang Quang, who was sitting on the edge of the runway. His legs were folded up against his chest with his arms wrapped around his knees. He was motionless and very quiet as I walked up to him.

"Mang Quang!"

I smiled and tried to appear positive and enthusiastic while he struggled to force a smile in return. There was little question that we were both concerned about what we would find upon reaching the village.

The night before, when visiting with some of the other Montagnards, they told appalling stories of things the VC and NVA had done to them over the years. So, as positive as I attempted to be, I truly feared for the safety of Mang Quang's family, and he appeared to fear the worst. Waiting gave both of us too much time to ponder his family's fate. We were more than prepared for the choppers to return so we could get on with our part of the mission.

Patting Mang on the back, I left him and started walking up and down and back and forth on the runway. Now and then, I would look toward the western horizon and glance at my watch. One time, only two minutes had passed—it seemed like fifteen.

Finally, the flight of helicopters could be heard approaching from the west. We immediately got our troops ready to go. As quickly as the Rat Pack ships settled in on the runway, troops began loading. In under five minutes, we were on our way back to the village.

Upon reaching the mountainside, I could see Captain Suber's chopper circling where Thieu Ta and I had been the day before. When we came in, we passed a little north and slightly downhill from the original LZ. Sullivan called as we passed by and said they had encountered no problems and had secured the LZ. "A few villagers have already walked in," he said.

The news was welcomed and immediately given to Mang Quang. With word of villagers arriving at the cornfield, life seemed to come back to his face. We both knew it meant the VC hadn't killed everyone after we left the day before.

When we neared the cornfield, which was overgrown with tall elephant grass, I pointed it out to the pilot. "Put us in there," I said.

As we turned on approach to what was to become our new LZ, the Rat Pack pilot asked, "Zero Two, do you want to light it up?" He was asking if his door gunners should fire on the field, if for nothing more than intimidation.

"It couldn't hurt," I said, "Go ahead."

With that, the command to fire was given and the door gunner on the side facing the field opened up, pouring M-60 machine-gun rounds across the thick stand of elephant grass.

Because there were stumps scattered throughout the tall grass, the choppers couldn't land. So, when we were as low as we could get, I jumped and Mang Quang came out right behind me. The draft created by the beating chopper blades whipped the grass and dried corn stalks in every direction. I squinted my eyes and moved out about ten meters to direct the off-loading troops toward my Vietnamese counterpart, who began placing them in perimeter positions on our new LZ.

The chopper lifted out and circled away from us. I moved farther into the field to act as a marker for approaching choppers while my Vietnamese counterpart continued to direct his men around the perimeter.

Almost every copilot on the incoming choppers gave some kind of positive gesture before departing. The copilot on the last one mimed two words that couldn't be heard over the roar of his chopper. But, even with the grass and dirt blowing everywhere, I could see well enough to clearly understand what he was saying. His message was simple, "Good luck." Acknowledging my understanding, I showed him a thumbs-up and nodded. His gesture of encouragement was the last I saw of him as his chopper passed over me.

When we started our insertion on the new LZ, Captain Suber, who was circling high above with Major Lee aboard, immediately shifted Captain Ted Dolloff and the Wolf Pack to cover us. Since he had commanded the Wolf Pack on the first mission, Captain Dolloff quickly led the Wolf Pack down the ridge to where we were about to start into the jungle.

When I looked up and noticed the wolves coming down the ridge, I felt great comfort.

Once everyone was on the ground, we pulled our perimeter in and moved directly into the jungle and toward the village. Sergeant Giao Phan, one of Thieu Ta Ngoc's most experienced platoon leaders, would lead the point unit. Mang Quang and I, along with my Vietnamese counterpart, were near the front of the column. The other American adviser and his radioman were toward the rear of the column.

According to Mang, in order to reach the village, we would have to cross only one small valley, a stream, and go up the ridge. I asked him to take us into the village by a route the VC would least expect us to be using. He said he knew a way, but it would be difficult.

He wasn't wrong. The jungle growth was so thick at one point that we had to get down on our hands and knees and crawl through it for some distance. Finally, we could see the sky again when we reached a small stream at the bottom of a ravine.

This is how some of the undergrowth looked as we began making our way up to the village. You can see one of the Vietnamese point team going into the jungle on the right. And, if you look closely you can see two others on the left.

The water in the stream was chest-deep. It had been stirred and muddied by the eight to ten members of the point unit ahead of me. When I reached the far side, I realized that more than mud had been stirred up by our crossing. There were nearly a dozen leeches on my forearms. They were disgusting to look at, but no reason to slow down. Pulling myself up into the jungle, I flicked them off and kept moving.

I had taken only a few steps into the thick growth on the other side of the stream when the hair on the back of my neck began to stand straight up. We were walking on a trail that had been very well-traveled. It appeared to be the jungle equivalent of an interstate highway. I signaled up the line to stop the point unit. When I turned back to see where Mang Quang was, he was just coming up out of the stream with Ahat and Light, my interpreter, close behind him.

When they reached me, I put my hand on Light's shoulder and spoke quietly.

"Ask Mang if we are close to the village."

Mang spoke in a whisper and gestured ahead up the trail on which we were standing. A translation didn't seem necessary, and there was no surprise when it came in a whisper as well.

"Trung Uy, he says we are very close. The village is at the top of the hill. He says this trail will take us straight up to the village."

I asked Light to pass Mang Quang's information up to the point unit and tell them to be alert. The same word was then passed to those behind us.

After enough time had passed for the message to reach the point unit, the signal was given for them to begin moving again.

As we moved up the trail, the jungle opened a little and wasn't quite so thick. I hated being on the trail. Since first entering the service, I'd been taught to stay off trails whenever possible. They're great places for booby traps and ambushes. But, searching both sides of the trail for signs of either, I noticed several large flowers standing out against the vivid green of the jungle. The setting seemed to contradict any concern about booby traps or ambushes.

No wonder the Montagnards picked this place to build their village, I thought. *It is beautiful; this could be paradise.* The thought had barely dissolved and I had taken only about ten steps when Boom! — Pow! Pow! Pow! —paradise exploded and gunfire ripped through the jungle.

Instinctively, I dropped to the ground and waited to see if the jungle around me was going to erupt with the gunfire of an ambush. When it didn't, I got to my knees and tried to look up the trail, but because of a bend, I could see only as far as two men—about ten meters.

It's the point unit. They've made contact, I thought.

In addition to the sound of the point unit's M-16s, the distinctive blooping sound of their M-79 grenade launchers could be heard. That was followed by the sound of impact farther up the trail. The explosions and gunfire up ahead were ear-splitting, and they were close. Shredded pieces of jungle foliage fell around me.

Later, thinking about the first few seconds of the incident on the trail, I considered some of the remarkable things that happen to the human body and mind during a crisis.

Science has demonstrated that the instinctive will to survive causes various body organs to rise to full function. Lungs work like billows as respiration increases, and the heart pounds in the chest as it pumps blood to other body parts demanding more oxygen.

While it flows undetected, the effects of adrenaline released into the bloodstream are unmistakable. Just prior to the contact, I remembered feeling a little tired because of our rapid and rigorous trek through the jungle, not to mention a lack of sleep during the past couple of nights. But once the gunfire erupted and I crouched on the trail trying to determine the seriousness of our situation, a tremendous feeling of energy surged. I suspect that it was simply the result of my brain preparing my body for a fight-or-flight response in order to survive. Since flight wasn't a possibility—I prepared to fight.

Despite all the things happening inside and around me, spontaneously everything seemed to merge and focus. I thought about what actions to take.

Gunfire was still coming from up ahead. I had no way to help or give intelligent direction without knowing what was happening, and the only way to know what was happening was to go up the hill and find out. I jumped to my feet and started up the trail.

Moving around the two or three Vietnamese soldiers who were directly in front of me, I began making my way toward the front of the column. Maneuvering up the trail, my eyes focused ahead and snapped alternately back and forth. With my thumb, I flipped the safety on my M-16 to the firing position and watched the jungle for the enemy.

At the same time, in anticipation of having an enemy soldier jump out somewhere in front of me, I kept thinking—*Don't shoot a villager by mistake!*

After reaching the turn in the trail, I looked back. Mang Quang and Light were right behind me. Then, turning back to go around the curve, I slammed into one of our Vietnamese soldiers who was running back down the trail. For a split second, I thought he was a VC and almost shot him just before slipping to the ground.

We were on a steep incline, and the trail was wet and slippery because water from a spring farther up the hill was running across it. Grabbing at plants, small trees, branches, and anything else I could get my hands on, I struggled to pull myself up the incline. But, as I tried to move up, other Vietnamese soldiers slammed into me. They were from the point unit and were hurrying to get back down the trail. When they reached the wet area, they used it as a slide and came down on their backs and behinds, often knocking me onto mine. Two or three times, I was knocked off my feet and was covered with mud by the time I moved past the wet area.

Finally reaching dry ground, I continued moving carefully up the steep trail. None of our men were anywhere to be seen, but I could still hear occasional gunfire. Not all of our point unit had come back down the trail, which meant Giao and some of them were still somewhere up ahead.

I kept moving and hoped that when I found them, they weren't jumpy and wouldn't shoot me by mistake when they saw me come up behind them.

As we worked our way up the trail, the radio Ahat was carrying became active.

"Zero Two, this is Zero Eight. What's your status? Over."

Sullivan had heard the gunfire and was trying to reach us to find out what was happening. Still huffing and puffing and trying to determine what was happening myself, there wasn't a time to stop to talk. But as we kept going, I turned around and pulled the handset off of Ahat's web gear.

"Eight, this is Two. We're going uphill and I don't know

what's happened yet. Over."

Because we were under dense jungle and on the opposite side of the ridge from him, Mike couldn't hear my transmission clearly.

He asked me to repeat, "Zero Two, can you say again? Over."

"Roger, Eight. Something is going on, but I'm not sure what. Over."

He still couldn't understand. "Two, speak slowly. I can't read you."

I knew he was trying to determine whether or not we needed help, but there was no time to attempt further communication with him. Speaking slowly enough for him to understand, I said, "Stand by, Eight. Zero Two, out." Then I passed the handset to Light.

Just about that time, a couple of explosions occurred that sounded like regular grenades or enemy rockets, and they were within meters. There was no way to determine who was shooting at whom or what, and I still couldn't see anything. Glancing back down the trail, only Mang Quang, Ahat, and Light were still with me—my faithful companions.

Farther down, it appeared that the Vietnamese had regrouped and were trying to follow us back up. Hopefully, that was the case.

Just as we reached a sharp turn in the trail, Mang Quang reached up and grabbed me and we stopped. There was no gunfire, not a sound. The haze of smoke and smell of gunpowder filled the air, so I knew we must be right on top of whatever had happened.

Mang whispered so low to Light that I couldn't hear what he said. Light leaned over to me, and while gesturing around the turn, he whispered.

"Mang Quang says that if you look around this turn you will see his village."

I looked at Mang and shook my head up and down. Then, I

moved in close against a tree and leaned around the turn in the trail. There they were.

Sergeant Giao and about five or six of his more experienced men were just ahead. They were squatting on the trail just inside the tree line.

One of the soldiers protecting their rear saw me and, to my relief, he recognized me. He tapped Giao on the shoulder and pointed toward me. When Giao saw me, he motioned for me to come up with him. My trio of shadows, Mang, Ahat, and Light, followed me around the turn and up the trail to the place where Giao and his men had stopped.

When we reached Giao, I whispered a greeting to him in Vietnamese. "*Ciao,* Trung Si."

"*Ciao,* Trung Uy."

Giao could not speak English well, so, in Vietnamese, he asked Light where the rest of his men were. Light told him that we had passed some going down, but that he thought they were coming back up.

Giao was angered that some of his men had run back and had not yet returned. He turned to the soldier who had seen me approaching and told him to go back and bring them and the rest of the unit up quickly. He told Light to tell me that when they arrived, we would secure the village.

Watching Giao's man going down the trail, I noticed the others coming around the turn. I tapped him on the shoulder and pointed toward them. Then, I pointed at the two of us and into the village. He nodded his head up and down.

Simultaneously we stood up and began moving from the tree line into the village. As we moved in, I ran my thumb along the safety on my M-16 to make sure it was still in the "Off" position and ready to fire.

The village was compact. There were six open huts built very close together under the dense growth of jungle. As we moved through the huts searching for someone who looked like they belonged to Mang Quang, I turned quickly this way and that. My index finger rested securely against the round, smooth curvature of the trigger.

I kept expecting either a VC or Montagnard villager to jump out in front of me and hoped there would be time to determine quickly enough which was which. I didn't want to kill a villager and didn't want a VC or NVA soldier to kill me.

The corner of one of the huts housed something that looked like a storage area. I thought I heard movement coming from inside and began moving toward it.

Regrouped, Giao's men were now pouring into the village, and there was a fair amount of chatter as the Vietnamese coordinated their movement into and through the village. I wasn't sure if I had really heard something or if my imagination was being creative. It could easily have been the sound of my own heartbeat pounding in my ears.

Very carefully, I reached for the cover to the storage area. Just as I snatched it open, there was a tremendous explosion right behind me. Dust and an assortment of debris blew everywhere.

Many times, I wished my ability to speak and understand Vietnamese was fluent. I understood enough to know that amidst the occasional gunfire and yelling, there was concern about a hole in the ground that someone else had discovered.

Light had taken cover under a table not far away, so I yelled to him.

"Light, what in the hell is going on?"

"Some soldier is saying he saw a VC go in a hole. He threw a grenade in the hole."

"What hole?"

"A hole in the ground . . . over there," he said, pointing as he spoke.

I rolled around to see where he was pointing but could only see some large rocks.

Getting up on my knees to take a better look, I could see holes around the base of a huge rock outcropping.

With my rifle pointed toward the holes, I stood up and moved over to see if there was anything in them. As I moved around looking in each one, there was nothing and no one to be seen, but they appeared to go very deep. I told Light to ask Giao to have a couple of his men stay behind to keep an eye on the holes while we finished searching the village.

As it turned out, there was nothing in the storage area I had been about to search when someone tried to crack the earth in half. But worse, there was no sign of Mang's family or any of the other families. The only people anyone had seen was a small group of VCs who were surprised by Giao and his men when they came up the trail toward the village.

After a brief exchange of gunfire and grenades, the VC had quickly disappeared down one of the many trails that ran off in various directions away from the village. As I checked one of the trails near where I was standing, I noticed movement in the jungle. From my vantage point, I had a relatively unobstructed view for some distance up the trail. The movement was perpendicular to and toward the trail. If the movement was a VC trying to exit the area, I was about to have a very clear shot. Raising my M-16 shoulder high, I leveled it up the center of the trail, again made sure the safety was off with my thumb, put my finger on the trigger, and waited for whatever was about to cross the trail.

Years before, while on one of our rabbit hunting trips, my grandfather had offered advice on what to do in exactly this situation.

"If you see a rabbit running toward the trail, wait till he gets there. Then, as he crosses it, take your shot and make a clean kill. Not even a rabbit needs to suffer," my grandfather said.

Standing positioned in the jungle foliage along the edge of the trail in order not to be seen, I waited for my shot as my quick-moving target approached. A small beam of filtered sunlight had made its way through the thick jungle canopy and was shining in one of my eyes, so I moved slightly. The nearer the movement got to the trail, the more the pressure on my trigger finger increased. The jungle was hot and steamy, and insects buzzed around my face. Perspiration was dripping down my forehead. *I hope it doesn't get in my eyes,* I thought. *As quickly as he's moving, I'm only going to have one shot.* Then, at that instant, my target burst from the jungle, thrusting back leaves and vines. *Fire, fire, fire!!!* was the screaming instinct to protect myself, but my eyes drew wide open. In a space of time too infinitesimal to measure, my finger straightened and came clear of the trigger. Standing dead center in my sites was a small Montagnard boy who quickly dashed off the trail and back into the jungle on the other side. In a long slow sigh, I released the deep breath I had been holding in anticipation of my shot and thanked God that I hadn't killed the child.

Later, we learned that the boy had become separated from his parents who were on their way to the cornfield, and he was simply trying to catch up with his family when he crossed the trail. He was slowed by the heavy load he was carrying, which included a large woven basket that included two of the family's chickens—one black and one white.

Back in the small village, I rejoined Mang Quang, who was upset that we had not found his family in or around the village. He expected them to be hiding somewhere nearby if they were still alive. Through Light, I told him we would continue our search and reminded him that several villagers had already reached the high cornfield and were safe. Maybe his family was there, too. I prayed that was true.

It was a bit gruesome, but I also asked Light to tell him to look around. There were no bodies and no blood. "No one has been killed here," I said.

Mang looked around, then shook his head and agreed that was a good sign.

I had just finished encouraging Mang when the rear part of our column arrived in the village. The American adviser who had been with our rear element walked over and wanted to know what he had missed.

I was giving him a quick rundown on what had happened when Lee called from Captain Suber's chopper to check on the operation. They were passing almost directly overhead. The jungle was so thick that we could hear the chopper but couldn't see it. Lee was checking to see if we needed anything after Sullivan had called him to report the gunfire and explosions he had heard.

"Bunkhouse Zero Two, this is Bunkhouse Zero Six. Over."

"Roger, this is Two. Go ahead, Six."

"Is everything okay down there? Eight said he thought you were in contact. Over."

"Yes, we were briefly, but everything is fine now. We're getting ready to move on toward Eight's location. Over."

"Okay, these guys need to refuel. I'm going to have them drop me off on the way in. Good luck! Bunkhouse Zero Six, out."

"Roger. Thank you. Two, out."

Shortly after Lee signed off, Captain Suber passed low over the treetops above us and checked in with me to confirm that he needed to refuel.

"Zero Two, this is Rat Pack Zero Six. We are going to drop your Six off, then head in to refuel. Stay safe. After refueling, we will standby at your base. Let me know when you're ready for us."

We rested for a few minutes while I finished briefing my team member on what he missed and what we were going to do next. Then, I found Sergeant Giao and told him we needed to move on up the ridge to the other field.

Giao shouted a few orders, and we were once again on our way.

Since we knew several NVA and/or VC with some significant weapons were still around, we left the village on one of the trails but quickly moved back into the jungle to avoid providing them with an ambush opportunity.

As our column chugged through the undergrowth like a slow-moving centipede, the sound of Tracer's engines roared and the speakers blared down into the jungle each time he passed over us just above the treetops. It was surprising to hear how well the tape could be heard through the jungle canopy. It made me feel better about the prospects of Mang's family hearing it and knowing they must reach the cornfield. Hopefully, we would find them before we got there.

I was just getting ready to push a thick growth of leaves aside when a hand suddenly appeared in front of my face, the hand of the soldier in front of me. Our movement had halted.

Motioning for Mang, Ahat, and Light to follow, I made my way up the column to see why we had stopped. When we reached the front of the point unit, Giao was standing next to a Montagnard family. Turning quickly to Mang Quang, I waited for a reaction, but it wasn't his family. It was, however, a family to which Mang was close because he was obviously very glad to see them.

The small family had heavy-looking baskets that were filled with their personal belongings. They were trying to carry them up to the cornfield as Tracer's tape directed.

The baskets were far too heavy for the children, though, and they were having a difficult time trying to move them. They had stopped to rest when Giao and our unit happened onto them.

When Mang Quang asked if they had seen any of his family, they told him, "No." They said they hadn't seen any of his family since the afternoon before. They also said that when Mang's voice told them to go to the cornfield, everyone ran to gather what they could and then started toward the field. For many with children and older family members, it was a long way and everyone couldn't make the trip in time before our unit had left. Those who hadn't reached the field in time had gone back into the jungle to hide from their enemy.

The place where the family stopped to rest wasn't far from the cornfield where Sullivan was waiting. It was a good time to let him know we were nearby and would be coming out of the jungle soon. I didn't want any of his team to mistake us for a VC unit and start shooting because I had long since learned that—friendly-fire was as deadly as enemy fire.

While alerting Sullivan to our pending arrival, I asked if any more villagers had arrived. "Yes, about twenty more," he said. Because we hadn't found them, I could only hope that Mang Quang's family was among them. Before continuing, we distributed the family's baskets among our soldiers and, once more, started up the ridge. It took only about fifteen minutes before we popped out of the jungle and onto the base of the cornfield.

As we moved farther up into the cornfield where it was a little more level, the Montagnards could be seen gathered together near the center of the field. I scanned the gathering, looking for a family that appeared to be missing a father. There didn't seem to be anyone in the group of villagers who matched the description Mang had given of his family. I was terribly afraid we were about to experience the same heart-sickening disappointment of the day before.

Then, deep into the crowd, I spotted a woman and two children sitting on a large rock. Turning to Mang Quang, who was also looking for the faces of his family, I put my hand on his shoulder and pulled him toward me. Then, pointing into the moving group toward the woman and children, I asked, "Mang Quang's?"

He looked and tried to see where I was pointing. For just a moment the crowd in front of the woman and children separated enough for him to catch sight of the three. When he turned back, there was no need for him to answer. Tears were again streaming down his cheeks, but this time they rolled over a smile that covered his face.

"Yeah, Trung Uy . . . Mang's," he, said patting himself on his chest. Gently, I patted him on his back and said, "Go!"

Mang had covered about half the distance to where his family was perched on the rocks when they saw him coming. One of the children yelled something in his tribal language that must surely have been the equivalent of "Dad" or "Daddy" in English. Then, the mother taking the smallest child in her arms, they jumped off the rock and hurried to meet Mang Quang.

I don't believe I've ever seen a happier person in my life than when Mang Quang was reunited with his family. The reunion was warm and emotional, and it was a special privilege just to witness. Mang had expected the worst possible outcome for his family but was experiencing the very best.

We were all relieved that the man who was responsible for our rescue attempt had been rewarded for his leadership and bravery with the greatest prize he could have been given, the safe return of his family.

Walking past them over to a flat place on the cornfield, I dropped my equipment and sat down on the ground. It was nice not to be walking uphill anymore. I looked out over the valley and could see Tracer still out there broadcasting. The two of Captain Dolloff's Wolf Pack gunships on station were making wide passes around the area, still looking for potential targets. Shading my eyes from the sun with one hand, I looked off in the distance. Two other choppers were coming in our direction.

"That's Rat Pack Zero Six and the third gunship coming back from refueling. They just radioed." Sullivan had come up behind me.

"That's what I thought," I said.

Gesturing toward Mang, he asked, "Is that his family?"

"Yes. They look glad to see each other, don't they?"

"Yes, they do," he said.

"How long has it been since the last villager walked in?" I asked.

"About forty-five minutes."

"Okay, let's have someone ask the villagers if they know of anyone else who is missing. Depending on the responses we get, we'll decide how much longer to stay. We'll be sitting ducks here if we hang around long enough for the bad guys to move in on us. And, if the NVA unit that's due to arrive shows up while we're here, we'll really have a problem."

"Okay, we'll check now," Sullivan said.

I took a drink from my canteen and got up to walk around and take a count of the Montagnards.

Most of the villagers were sitting as I wandered through the group counting and smiling. Many seemed pleased to be where they were. Others seemed frightened and uncertain about their future. A few seemed suspicious as if believing they might simply have traded one set of captors for another.

Light walked up with a bullhorn and asked the Montagnards if anyone knew other people who were missing. A few names were called out, but Mang Quang said he had seen them at Trung Dung. They had been picked up the day before.

Eventually, after several other names were discussed, we realized no one really knew exactly how many villagers were still in the area. We had been told that the villagers had been separated and denied any frequent contact for purposes of control. So, it seemed the VC's strategy of separating the Montagnards into the smaller groups had been effective.

Over the years, since they had been kept from communicating freely, the villagers had become uncertain of their own numbers. For that reason, it would have been difficult if not impossible for the unarmed Montagnards to organize any successful attack against their captors. Also, Mang had told us that he knew some of the Montagnards had run away and fled deeper into the jungle to escape VC control. It became clear that until we got the villagers in one place to make some kind of list, we weren't going to know whether or not we had everyone.

If I called for a pick-up at that moment, it would still take almost two hours to get everyone out. That would allow additional time for any straggling Montagnards to reach us. It was late in the afternoon and rain was moving in from the coast. It was time to begin our extraction. We would do our accounting of the villagers at Trung Dung.

I radioed Captain Suber, who was on standby back at Trung Dung with Ngoc and our reinforcements.

"Rat Pack Zero Six, this is Zero Two. Over."

"Roger, this is Rat Pack Zero Six. Go ahead. Over."

"Come get us and bring us home. Over."

"Roger, we're on our way. Over."

"Thank you, we'll be standing by. Out."

A couple of minutes later, Captain Suber called back. "Zero Two, Rat Pack Zero Six, over."

"Roger. Go, Zero Six."

"We are inbound. ETA approximately two-five mics (minutes). Over."

"Roger, we'll be ready."

"Zero Two, Zero Six. I've got some guys who want to know if our man (Mang Quang) found his family."

"Roger, Zero Six . . . he did. They're standing here with him right now. Over."

"Great! I'll pass the word. Makes you feel really good, doesn't it?"

"It certainly does," I said.

No other Montagnards appeared on our cornfield LZ before the last chopper dropped in to extract me and the last six remaining Vietnamese troops. Until that chopper cleared the LZ, Dolloff, and the ever-watchful Wolf Pack pilots and crews circled overhead providing protection.

We had recovered another thirty-six villagers who had made their way to the cornfield, and now, the last of them were headed for safety. And, this time, we knew we had Mang Quang's family.

For a second time, spirits were high as we flew back to Trung Dung, but they became even higher when we reached camp and the two groups of Montagnards were reunited. In all, we had recovered one hundred-eighteen villagers. Joy created by the long-overdue reunion of the village immediately caused a spontaneous celebration.

While it seemed of little importance to the Montagnards who reveled in their new taste of freedom, everyone who took part in the mission knew we had been far more fortunate than we had any reason to expect. We had taken no casualties in only minor skirmishes in a place far outside of our normal area of operations.

Now, despite any risks that may have been faced, the men of A-502 offered the Montagnards more than their exceptional military skills.

As the villagers continued to greet each other as if long-separated relatives, I watched our team members distribute food, drink, and bedding and marveled at the respect and kindness with which these seasoned military men treated the primitive families. With warm smiles and unexpected gentleness, some even helped make beds for smaller children who had obviously been made weary by their experience. These scenes weren't being filmed or reported to anyone back home, which made their occurrence seem so much more sincere and genuine.

By going where they had gone, the men from A-502, the Bandits, the Rat Pack, and the Wolf Pack from the 281st, along with Tracer from the 8th Psy Ops had all risked a great deal—the ultimate. Watching as some of them shared in the joy of the moment, I wondered, *how could anyone ask more of those men?* But, before the week was over, that's exactly what I would be doing. However, this time, would-be rescuers would find themselves in need of rescue and in peril of the worst possible fate for a soldier.

More Families Ferried to Freedom

Villagers gathering for count and listing, willingly followed direction and helped as we tried to create a list of families.

Uncertain and Apprehensive

While they acted glad to be freed from slavery, a few of the Montagnards also seemed uncertain and apprehensive about their future. Team members smiled broadly as they moved through the villagers trying to reassure them that they had reached safety and caring hands.

Class Picture — Rescued and Rescuers

After the last group of villagers had been safely landed, someone organized this and other pictures. Thieu ta (white t-shirt) and many members of the 281st with some of the freed villagers

The best part of this picture for me has always been seeing Mang Quang holding his son. That was a very special day.

Welcome to Freedom

A-502's Commanding Officer and Senior U.S. adviser, Major Will Lee is welcoming arriving Montagnard families to Trung Dung.

A New Mom and Her Newborn Baby

This baby was born the day before our rescue team arrived. The young mother was happy to let our medics check and clean her baby for her. She walked and carried her baby through the jungle barefoot, just as she's dressed, to reach the cornfield.

Vietnamese Airborne Special Forces Patch

When we finished going over final mission details the night before the first rescue mission, Thieu ta, the man I would come to think of as a brother, gave me one of these Vietnamese Special Forces patches. As he handed the patch to me, he said that it would bring me luck and protect me. Since it would hang from a buttonhole and hang over my heart, I asked him if it would stop a bullet. He laughed and said, "Yeah, sure!"

While doubting its power as a shield, I accepted the patch with great pride. The patch can be seen hanging from my pocket in the picture on the next page and when I was interviewed by David Culhane of CBS News during the rescue.

Sadly, sometime during my return to the United States, the patch and other items were taken from my baggage. Although little more than a stitched piece of cloth, the patch had great personal meaning to me because it had been given to me by Ngoc. And, because it was a cherished symbol of the rescue that my friend Ngoc had made possible.

In 2019, more than 50 years after both presentation of the patch and the rescue, Ngoc's oldest son, Chau, learned that my patch had been taken. To my great surprise, he had it reproduced and sent it to me as a gift. Now, the red tiger patch has even more meaning to me.

Tom, Mang Quang, and Interpreter on the LZ

Wearing the "bulletproof" Vietnamese Special Forces "Tiger Patch" given to me by Thieu ta, I was preparing to give Mang Quang the task of calling missing villagers to the cornfield. The third man is my interpreter.

Remarkably, even though his family was safe at Trung Dung, Mang Quang insisted on being at my side when the rescue team attempted to find and free the last of the missing villagers.

When this picture was taken by one of the reporters, the three of us were watching A-502 team members set up a large broadcast speaker that Mang Quang would use to call his fellow villagers to freedom.

CHAPTER 15

Ice Cream and Little People

AFTER WATCHING THE MONTAGNARD reunion for a while from a distance, I went to the team house and asked one of the Vietnamese cooks to put ice into tubs of soft drinks. When they were ready, he helped me load them into the back of one of the jeeps. We then drove the cold refreshments down to the tent, where the Montagnards were introduced to Coca-Cola.

While the villagers began to relax in their new surroundings, Sergeant Jerry Arrants and his team of medics moved through the new group doing checkups. They took their time, administering medicine, applying Band-Aids, and kindness as necessary.

Other team members arrived and passed out clothing they had gathered from here and there while still others helped distribute food that had been prepared for the many hungry families.

Sergeant Koch went to one corner of the tent and began the task of creating a list of everyone who had been rescued. When it was finished, we would visit with each of the adult Montagnards to determine how many, if any, were still missing.

While almost everyone else on our team was busy healing, clothing, and feeding, I took one of the cold Cokes, walked about twenty meters away, and crawled up on a stack of sandbags. Leaning back into the bags, I looked over toward the tent where 123 Montagnard villagers were being cared for by a group of Americans who acted more like hosts at a party than soldiers.

As I sipped my Coke, I found great satisfaction in what had occurred during the past two days. And, particular satisfaction in what was continuing to happen at Trung Dung. With the ongoing scene of warm human interaction before me, I wondered who was benefiting most from the experience—the Montagnards or our team members. For a few moments, my mind yielded to things philosophical, probably not a unique occurrence during a time of war.

The events of the past two days were a side of war one rarely if ever, has the opportunity to see. *I am fortunate to witness this,* I thought. The crying, pain, wounds, fear of death, and death itself that the Montagnards said they had experienced in the past were, for the moment, replaced with laughter, healing, caring, good food, and the anticipation of a new life.

From my vantage point on the sandbags, I watched Lee, Lane, Vasquez, Phalen, Arrants, Freedman, Sotello, Trujillo, Stewart, Koch, King, Bardsley, Key, Miller, Lavaud, Hawley, and all the others as they came and went with individual attempts to make the Montagnards feel safe and welcomed. Watching them, I mused at how different the Americans were. They were black, white, and brown. They were Jewish, Christian, and who knows what else. Despite their differences in race or religion, they functioned as the team they were and with a single purpose as they cared for the Montagnards. They were doing far more than the Special Forces

Motto, Free the Oppressed, suggested. These combat seasoned soldiers, who some might consider hardened, were now caring for villagers—as if they were family.

What an interesting phenomenon it is, I thought as I watched my teammates, *how human beings of all backgrounds bond together during war or other times of trouble or disaster.* Differences in skin color, religion, and politics seem to become less important during difficult and challenging times as focus shifts from differences to similarities.

When we had free time, team members would often relax in the team house and talk about home. We talked about things we liked and things we didn't like. During one of those times, I realized how distinctly different the men who made up the A-502 team truly were. Yet, they worked incredibly well together and seemed to care very much about each other's welfare. Our team represented a microcosm of the country from which we had all come.

Watching as another team member arrived with more blankets to be used as bedding for the Montagnards, it occurred to me what an extraordinary amalgam of humankind we Americans truly are. We are woven together into a wonderfully vibrant fabric from the threads of countless ethnic origins. Even though a blend of differences, we all share something that gives the fabric its amazing strength. Americans, like no others, share an unyielding determination to live in freedom.

Still watching while the Montagnards enjoyed their first day of freedom in many years, I witnessed firsthand the joy of living free. Before this day, freedom was just a word to me, a concept. It was a dream that brought the Pilgrims and other settlers to America hundreds of years ago and continues to bring them today. But now, freedom was more than a word, a concept, or even a dream.

Freedom was something real—I could see it and I could feel it.

I have already mentioned that my father flew with the 100th Bomb Group of the 8th Air Force during WWII and his B-17 was shot down by the Germans. Fortunately for both of us, he survived the crash landing. Though the experience of being blown out of the sky had been a harrowing one, my Dad said that he never regretted his service. He said that World War II was a fight for the freedom of millions of people. He felt that effort to perpetuate and ensure the dream of freedom was one of the many things that made the United States great and gained our country recognition around the world as the champion of human rights.

While I had always remembered my Dad's sentiments, I also realized that, as a nation, we are not perfect and we do occasionally make mistakes. After almost nine months in Vietnam, I had grown in maturity regarding political issues and was beginning to be concerned about our involvement in the war. It would not have been difficult for me to challenge Lord Tennyson's postulation that it is "Ours not to reason why—Ours but to do and die." However, as I watched my fellow team members tend to the Montagnards, I felt reasonably sure of one thing. Whenever the United States offers to share our dream of freedom with others, men such as these would surely do their very best, as they had this day, to ensure that the dream became reality.

My momentary philosophical drift was jolted back to the here and now when Miller skidded up in a jeep.

"Hey, LT!"

We were often less than military formal at 502, so I returned Miller's spunky greeting.

"Hey, Miller. What's up!"

Looking toward the tent housing the Montagnards, he continued. "Looks like everybody did a great job today."

"Without question, they did an incredible job. I was just thinking about what everyone has been doing. The folks at home would be pretty proud of them. Don't you think?"

"Yes, sir, I do. And, you know I wanted to go, too."

"Yes . . . I know," I said, laughing as I spoke, "Everyone wanted to go."

Then, wondering what he had come for, I asked, "So, Miller . . . did you need something or was this just a driving demonstration?"

"Oh, yeah . . . No. Lieutenant Orians has been calling all afternoon to find out what happened. He's on the radio now. If you can come back with me . . . he'd like to talk to you."

"Sure. Let's go."

We drove the short distance to the team house, and I went to the radio room to give Frank a report on the day.

He was pleased that the villagers had been brought in, but was disappointed that immediate media coverage hadn't been available. He said he was going to alert the various in-country media bureaus for follow-up coverage and told me to expect visitors.

I told him we'd be ready and thanked him for checking on us.

After cleaning up a little, I went to report to Thieu ta, who had just returned from his meeting at LLDB High Command in Nha Trang.

He was sitting on his patio drinking a beer as he shuffled through notes from the meeting. When he saw me approaching, he shouted, "Trung uy . . . good to see you alive! I am glad the patch I gave you is working. Come have a beer and celebrate your mission with me."

He was referring to the red tiger patch he had given to me just before the mission and he had, obviously, already heard that the day had gone well. I was glad to join him and we talked more like friends than soldiers about the events of the day. Just before I left, Ngoc put his hand on my arm and said, "I can tell that you are happy . . . that makes me happy."

As I stood up to walk away, I said, "Thieu ta, this event has made my coming here worthwhile . . . and you made it possible. Thank you."

Then, I went back down to spend the evening with the Montagnards. When things grew quiet there, I finally went to get a good night's sleep.

The next day, as we began questioning the Montagnards regarding the list Sergeant Koch had compiled the night before, it became apparent almost immediately to both of us that other families were still missing. Several of the villagers we visited with told us that people they had seen recently were not among the group relocated to Trung Dung.

After spending most of the day with villagers and interpreters, the count was between forty and fifty people still unaccounted for, depending on who you talked to. We thought one or two people might have been referred to by different names, but that made little difference. We were reasonably sure that between forty and fifty people were still "out there." That was a significant number and, like it or not, we felt that we had to consider going back.

However, we discovered another serious problem. Based on information obtained during our questioning of the Montagnards, we began to believe that the enemy unit that had been using the village area as a way station was part of the 18B Regiment. It was a well-armed and very disciplined North Vietnamese Army unit. One of our units had faced elements of their 8th and 9th Battalions in the Dong Bo Mountains with catastrophic results. We postulated that it might be members of the 7th Battalion that was due to pass through the area of the village any day now. The 7th was away when the other two battalions were annihilated by the Koreans. So, it made sense that they would try to reestablish a presence somewhere in the Nha Trang area.

Regardless of who was using the village, we couldn't make repeated trips so far outside of our AO into the same area without running the risk of being ambushed or encountering a large NVA unit like the 7th Battalion. And, it didn't give me any comfort that, during our many visits, several of the Montagnards mentioned a unit that was expected soon.

After considering our dilemma, we decided it would be wise to let the VC and/or 18B NVA think we were finished and wouldn't be coming back. We decided we would let a few days pass before attempting to rescue the remaining families, but I immediately began making plans and arrangements for our final trip to the village.

When I spoke to the 281st, Tracer, Thieu Ta, and the other American advisers, there was no attempt to minimize the seriousness of the situation. Our predicament and the danger of returning a third time were made very clear, but no one flinched. Everyone agreed we couldn't just forget that the remaining families were out there. We would have to go after them and return to the village—one more time.

Sergeant Koch and I, along with South Vietnamese Intelligence, spent the days before our final trip questioning the villagers about the various enemy units that they had encountered during the past several years. We wanted to collect as much information as possible. The Montagnards reported that, besides those of 18B headed to the Dong Bos, many other enemy units passed through the village en route to other destinations.

During the questioning, we discovered the unit responsible for killing Dale Reich had very probably come from the village. While discussing some of their captors' most recent activities, two of the villagers mentioned that a few months before, a unit from 18B had returned to the village after being gone only four or five days. The villagers overheard them discussing an ambush their unit had walked into while they were away. We felt reasonably sure the unit they had encountered was ours and the one to which Dale Reich had been attached. The villagers said the unit had an unknown number of men killed and many wounded in that battle. They said two or three others died after returning to camp. The Montagnards were sure of this because they were made to dig the graves for those who died.

Team members quickly developed a desire for revenge when they learned that the enemy unit holding the Montagnards might also have been responsible for Dale Reich's death. While I understood and shared the feeling, I wondered if what we were doing might be the greatest moral revenge that we could take on Reich's behalf. The enemy unit had taken one life from us. We had already taken 123 from them and planned on taking another forty, or more. There was a difference, though. The ones we had taken would continue to live.

After draining the Montagnards of intelligence information, we turned to lighter things. I took Mang Quang's children and several of the others up to the team house where we dared to serve them chocolate ice cream. It's probably no surprise, but they had never tasted ice cream before. And, watching their reactions was great fun. Almost all of the children examined the brown concoction carefully and smelled it before attempting to eat it. Team members encouraged them by helping themselves to a bowl and demonstrating that it was for eating. Their demonstration included a variety of sounds and facial expressions meant to show how good ice cream was.

When the children finally began to eat it, some loved it immediately and couldn't get it into their tiny mouths quickly enough. Then, others tapped on it with their spoons or used their fingers to determine its exact nature before it reached their mouths.

By the time the ice cream was gone, the hands and face of every child were covered with chocolate. While our ice cream party required one heck of a well-supported mopping-up operation, the best was yet to come.

Coming from an industrialized country, we take so much for granted. Things that are commonplace to us may seem inconceivable or even fantastic to someone as primitive as the Montagnard children. They had all been born and had grown to their young ages in the jungle. While I thought they might enjoy it or be amused by it, I was not prepared for their reaction when I turned on the—television.

We seated some of the children on the sofa in our sitting area and the others on the floor in front of them. One of our team members had taken one of Pop's dishtowels and was teasing them by holding it in front of the television while we waited for the picture as the TV warmed up.

A rerun of an old sitcom was on, I don't remember which one, probably "I Love Lucy," but when the towel was popped away from in front of the screen, their reaction was unbelievable. They all simultaneously sprang to their feet and began screaming, pointing at the television, and running around. It was instant chaos!

Three of the older boys ran to the television and tried to look inside from the top, sides, and back. One of the boys tapped it on the side and seemed to be trying to talk to it. Then I realized what was happening. The children thought the characters on the screen were alive inside this magic box on the table.

To calm things down and regain control of our ice cream party gone wild, I reached over and turned the television off. Attempting to restore calm, team members made repeated shushing sounds. After everything became quiet, I went and found Light to help me explain the wonder of television to the children who we had just inadvertently yanked from near prehistoric times into the twentieth century.

The next day, we were still laughing about our attempt to entertain and amuse the Montagnard children. All of the Americans seemed to enjoy sharing their culture with the villagers, who were quick to return the gesture by sharing theirs with the team.

I'm not sure who had the toughest time adjusting to the other's customs. Some of the Montagnard women felt no need to cover their breasts in our presence. But, as unsettling as that was at first, we got used to it and they were all treated with respect and given shirts to wear.

Something equally difficult to adjust to was the sight of young, even tiny, children smoking. The first time I saw a child with a pipe I thought the small boy was a dwarf. Even after getting very close to him I still wasn't sure he was a child, but he was. The young boy, who was only nine or ten years old, smoked his pipe like he had been doing it for years. As I stared, he smiled, then offered me the pipe. I shook my head back and forth and coughed, indicating what would likely occur if I accepted. He just laughed and put the pipe back in his mouth and continued to puff. He seemed to enjoy his pipe, maybe more than our ice cream.

The more I was around the Montagnards, the more I noticed the smallest children that appeared to be no more than two, three, or four years old, smoking cigarettes. Obviously, their parents had never seen the Surgeon General's warning about the danger of smoking to your health. While amusing, I never got used to the sight, which was clearly a part of their culture that began very early in life.

Our ice cream party was one of the few, but very welcomed days of fun during the war. Days like those helped us become centered again, reminded each of us who we truly were, and conjured warm thoughts of home.

Differences in Culture

Even the CBS News crew found interest in the Montagnard children's smoking because they captured these images. The young boy in the picture above is the one who offered me his pipe later at Trung Dung.

CHAPTER 16

Who Called?

DURING MY TOUR OF duty, one of my most surprising experiences didn't occur in the jungles of Vietnam. It occurred within the relative security of the Trung Dung compound and was so unexpected that I still find it difficult to believe.

I was in the Operations office the day after our run-a-muck ice-cream party, working on plans for our return to the Montagnard village when I was interrupted. One of our radiomen poked his head in the office door and said, "LT, there's some guy from the Red Cross on the radio and he wants to speak to you."

"Okay, I'll be right there."

As I got up and headed to the radio room, my thoughts immediately turned to home. *Something has happened to someone at home*, I thought, I feared.

When a serviceman was killed in Vietnam, the family back home could be notified one of two ways. They could receive a very impersonal Western-Union telegram or by a more personal in-person visit from a member of the service, a chaplain when possible. Even though more considerate, that knock on the door was one no family welcomed. The use of the distasteful and the much-dreaded telegram was eventually discontinued.

In the reverse, it was the Red Cross that often made the notification to service members of the injury or death of a loved one back home. So, it was with serious concern that I entered the radio room.

When I acknowledged the call, an ominously calm voice on the other end of the radio identified himself by name and said he was with the American Red Cross. Then he asked, "Is this Lieutenant Thomas A. Ross of Pensacola, Florida?"

After saying, "Yes, it is," I prepared myself for whatever bad news he was about to deliver.

"First," he said, "I want you to know that everyone back home is just fine." Then, he told me why he was contacting me.

I couldn't believe what he said. So, to confirm what I had just heard, I asked, "Who called?"

"Your mother. I just hung up from speaking with her."

"What?"

"Yes, she called the office here in Saigon. This is a combat zone and civilian calls are not supposed to be routed here."

"Yes, I'm sure," I replied. "Unless it's my mother on the other end of the line. She's a very persuasive and determined woman."

"I know, she said someone in Washington had given her our phone number. She wanted me to put her through to you at A-502, but I told her I couldn't do that."

After laughing out loud, I asked, "So, what did my mother want?"

"She said she hadn't received mail from you in two weeks and wanted to know that you were okay."

Before the rescue developed, I had been out on a 10-day long-range patrol west of Trung Dung. I had written a letter home the day after we returned, but it hadn't had time to reach home yet.

Continuing, the man from the Red Cross said, "Your mother told me that she wouldn't hang up until I promised to call her back after confirming that you were okay. So, I did."

"Well, you are very kind. Please tell my mother that you spoke to me directly. That will give her comfort. Then, please tell her that I said I am just fine and that I love her very much. Oh, tell her that I asked about my Dad, my sister, Polly, and my dog, Bingo. That will confirm that you spoke to me and that news should make you her hero."

He laughed, then said, "I will. I'm going to call her right now."

I thought about whether or not to include this story in the book, but how could it be left out? Families experienced the war right along with those of us who fought it. And, maybe even more so—imagining all the terrible things that might be happening to us.

Recognizing the American Red Cross

During the time the American Red Cross was in Vietnam (1965-1972), its staff handled more than 2,168,000 emergency communications between servicemen and their families. And, my mother's call was one of those.

Military commanders welcomed the Red Cross's presence in Vietnam, saying that their services were "indispensable and prime factors" in maintaining high morale and caring for their men.

The year I was in Vietnam (1968), General William Westmoreland, then commander of the U.S. forces in Southeast Asia, wrote a letter to the Red Cross's national headquarters. The general commended the organization by writing, "Serving our men on the battlefields here in Vietnam, the American Red Cross is a hotline to the folks back home, an oasis in the heat of battle, and a comfort during hospitalization."

The Red Cross perfected its services during World War II, the Korean War, and all the other years it helped millions of servicemen and women cope with personal and family emergencies and hardships while serving their country and, often, far from home. In my case, I probably couldn't have been farther from home when my mother needed to know that I was still alive and well.

I offer my sincere "Thank you" to the Red Cross for its service to my family.

CHAPTER 17

One Last Time

DURING THE DAYS WE waited to return to the village one last time, my already developed affection for the Montagnards continued to grow. They were relatively quiet people and kept to themselves. They seemed comfortable in their temporary surroundings and, even though primitive, possessed the courtesy to express appreciation for the care that they received at Trung Dung.

Whenever wandering through the tent area where the Montagnards were housed, I often looked for Mang Quang's children. Frequently, before going to visit, I would go through the goody boxes sent from home by my mother to look for something they might enjoy. My entire supply was depleted in only three days because I also took things to share with other children who quickly learned to look at my pockets carefully when they saw me coming. When it came to identifying objects in my pockets, it amused me how short the learning curve was for children who had been born and raised in the jungle. They were able to detect a protruding candy wrapper or the outline of cookies at amazing distances and would show me big grins.

While partial to the children, Mang Quang's in particular, I had warm feelings for all of the Montagnards. To me, they had become a tangible symbol of why American soldiers had come to Vietnam. They were warm people who wanted nothing from anyone other than the opportunity to live peacefully in freedom in their small, generally isolated village communities.

Because of my affection for the villagers, a Vietnamese soldier who seemed jealous of all the care and attention the Montagnards were receiving from our American team members drew my ire. He walked up beside me with a question one afternoon during one of my many trips to visit the Montagnards, "Trung Uy . . . why Americans give so much help to Montagnards? They are *jungle* people," he said.

There was already reason not to like or trust this particular soldier and his question only served to increase my distaste for him. His attitude regarding the American advisers had often seemed negative, and I had once caught him sneaking around our team house. Since we already knew that our CIDG units included VC sympathizers, it had crossed my mind that he may be VC. Unfortunately, we hadn't yet proven our suspicions.

In response to his question, I looked straight into his eyes. "We help *them* for the same reason we came to help *you*," I said.

It is unlikely that my answer meant anything to him, but he didn't say anything else and simply walked away.

The morning before our third and final trip west to the Montagnard village, interest in the operation began to spread beyond Trung Dung and the walls of the Dien Khanh Citadel.

Frank Orians called to say that AP (Associated Press), UPI (United Press International), and other media agencies had contacted him, as had one of the three major television networks, CBS News.

In 1968, there were only three major television networks for news. CNN, Fox, and others were still decades away.

Frank said the news agencies were interested in the Montagnard rescue and would probably be out within the next day or so to cover the story. He was surprised, then enthusiastic, when I told him there might be even more to report because we were going back to the village one last time.

"Why?" he asked.

"In an attempt to retrieve families that are still missing," I said.

Even Frank, who was a Special Forces–trained PIO, recognized the seriousness of returning to the village a third time.

"Isn't that risky?" he asked.

"Yes, but what can we do, leave them out there? We're all agreed, even our support units. Leaving them out there is simply not an option."

"When are you going?" he asked.

"Tomorrow morning . . . as soon as there's enough light to see where we're going."

He wished me luck and said he would visit the camp, not later than the next afternoon to prepare a news release on the story. I told him we would be glad to see him and would introduce him to our new friends.

By later that afternoon, word of our return to the remote mountain village had spread around 5th Group Headquarters in Nha Trang. A lieutenant colonel from Operations called to say he would like to go with us the next morning. He said he had been behind a desk too long and wanted to get into the field. "This sounds like something I would very much like to be involved in. Do you mind if I tag along, Lieutenant?" he asked.

What was I going to say to a man who outranked me by several ranks? I told him we would be glad to have him along and suggested that he catch a ride out with the 281st the next morning.

Preparing my operations order and coordinating plans for the mission helped pass what was left of the day. Somewhere amid our planning session, Sergeant Roy King, my Operations Sergeant, who was capable of producing nearly anything you needed, presented me with a gift. It was a new plasticized map that included the Montagnard village area. "I don't want you to get lost," he said as he handed it to me.

"Where were you a week ago?" I asked.

Sergeant King laughed and said, "Just be glad you have it now in case you have to walk back."

I thanked him, not imagining that his words might prove prophetic.

It was evening by the time I finished plotting positions and making notes on my new map. I folded it so that the cornfield was in the center and headed for the team house. Having missed dinner, I went searching for something to eat in the kitchen. Upon opening the refrigerator door, I was surprised to see a plate with my name on it, LT ROSS. One of the cooks had covered a plate of food for me. I took it to my room and, after finishing it, had no problem falling asleep.

There was just a glimmer of daylight when I slung my web gear up over my shoulders and fastened it in place. The screen door at the end of the team house slapped shut behind me as it had many times before, and I started walking toward the airstrip. Behind me, the high-pitched hum of an approaching aircraft was immediately audible. Then in a second or two, Tracer zipped overhead, followed closely by the 281st. They were early. It seemed there would be as much enthusiasm about our third trip to recover Montagnard families as there had been for our first two trips.

I hadn't taken more than a few steps more when Thieu ta roared up beside me in his Jeep. "Good morning, Trung uy!" he bellowed as if he had been awake for hours.

"Get in!" he said, "I'll take you to the airstrip."

I jumped in and away we went. As we rode, Thieu ta glanced over and eyed the red tiger patch he had given me the night before our first trip to the village. It was hanging from my left pocket. "No bullet holes in your patch. Make sure you come back with it the same way tonight," he said.

Thieu ta was trying to be light-hearted, but he was concerned about attempting a third mission into such a remote location. I knew that because, the day before, he was leaving his intelligence office when he saw me crossing the compound and called out to me. He asked me to have lunch with him.

We had barely begun to enjoy our classic Vietnamese meal when he started to share some apprehensions based on what he had just been told by his intelligence team.

"Trung uy, I am concerned about you returning to the village of the Montagnards tomorrow," he said. Then continued, "My intelligence unit just told me that one of our agents has reported that a major unit of 18B is due to arrive in the Nha Trang area any day now. My concern is that the agent believes the unit will approach from the west and is expected to pass through the area of the Montagnard village."

"Yes, I know," I said. "While visiting with the Montagnards

they told me that, just before we first arrived at the village last week, they heard their overseers discussing the unit's return sometime soon. We believe it is probably the 7th Battalion."

"Then you realize how dangerous tomorrow might be for you?" Ngoc queried.

"Yes, I do. So, as on the last two missions, I have a pair of Air Force F-4s on standby in Cam Rahn Bay along with Spooky in Nha Trang. And, Tracer, the Air Force pilot flying the speaker plane, is prepared to take on the role of a Forward Air Controller to direct any needed air support. Also, I have a larger reinforcement unit on standby today."

"Good! All sounds good. But, Trung uy, you must remember that NVA soldiers are highly trained and are very dedicated to taking over South Vietnam. They will enjoy killing you if they are given the chance."

"Are you trying to scare me, Thieu ta?" I asked.

"No, I just want to make sure that this is something you want to do and you are prepared for what might happen tomorrow. I want to see you get off one of those helicopters tomorrow evening."

"Yes, Thieu ta . . . this is something I want to do. Knowing how they are being treated, I would be haunted for the rest of my life if we just left families out there. Every time I looked at mine, I would think of theirs. And, I can tell you that every member of A-502 supports the mission. They all want to go and we will all be well prepared."

"Very well. Then, I will be with the reinforcements at the airstrip all day tomorrow. If you call for reinforcements, I will lead them."

When we reached the airstrip, Captain Dolloff and the Wolf Pack had arrived first and were already on the ground. The slicks, either the Bandits or the Rat Pack (I didn't know which), were circling to land, and Tracer was now off in the distance making a turn.

All, Ready to Go — One Last Time

Fog hung thick over the Song Cai River, which was north of our airstrip, and morning mist filled the valleys as we prepared to begin our third and final mission to the Montagnard village.

Ready for a Battle

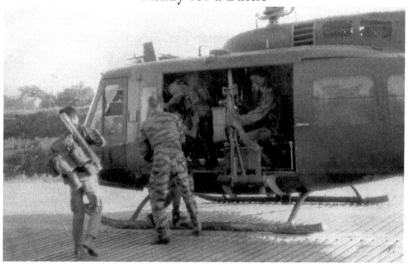

We were prepared for whatever might happen and carried more in weapons and troops than we had on either of the two previous missions. In case Thieu ta was correct and we did encounter an NVA unit, we were ready for battle.

Looking up the line of choppers, I saw a familiar figure approaching and was a little surprised to see him. It was Frank Orians, and he was in his combat gear. I waved and went to meet him.

When close enough for him to hear me, I asked, "What are you doing here?"

"Go look on the third chopper," he said. "I brought you a surprise."

"Oh, really?" I responded, having no idea what his surprise might be.

Following his direction, I walked up to see what he had brought. Upon reaching the chopper, I found it full of press representatives. As they exited the helicopter, Frank introduced me to David Culhane from CBS News, his cameraman, and his soundman. Then he introduced Don Tate, a Scripps-Howard war correspondent, and staff writer. There was also a man from another major print publication such as Look, Life, or Saturday Evening Post, I don't remember which. Unfortunately, I have forgotten his name, but not his kindness. Sometime after the rescue, he sent me a picture he had taken of me after our last trip to the village.

When the last person climbed down out of the chopper, I was surprised to see she was female. As she left the chopper, Frank introduced her, but because I was thinking, *Surely, she doesn't plan on going with us*—her name went in one ear and out the other.

I do remember that Frank said she was a photojournalist with UPI (United Press International). While I have forgotten her name, I will never forget my encounter with her. She taught me something very important about the determination and courage of women.

The young female reporter had a long-lens camera hanging

around her neck and a pad and pencil in one of her hands, with which she had already been taking notes. She was attractive, with short auburn-colored hair, bright, penetrating eyes, and appeared to be in her twenties. While her posture and conduct suggested self-confidence, she would soon demonstrate her assertiveness and her ability to represent herself forcefully.

After some initial chit-chat, I asked the female correspondent to step aside from the others with me. I pointed through the opening in the fort's wall and told her that the Montagnards were just beyond it and that A-502 team members would be there to help her with her story. "I'll visit with you when we get back," I said.

Peering at me intently with a quizzical look on her face, she responded firmly, "What do you mean? I've been cleared by 5th Group for this mission. I'm going *with* you."

"That's not possible," I quickly countered. "You're going to have to stay here until we return."

"Are they going?" she asked, gesturing toward Culhane, Tate, and the others.

"Yes, they are," I said.

"Then, why am I not going?" she demanded.

"I don't want you to be hurt," I said sincerely and with some feeling in my tone.

"Do you want them to be hurt!?" she demanded, again gesturing toward the male reporters.

"No, I don't want them to be hurt either," I replied, noting the fire in her now slightly squinted eyes.

"Good! Then, leave them here with me!"

She was, obviously, extremely upset, so I didn't laugh. But, the others, who were positioned behind her after she had turned to face me, either smirked or grimaced. Then they moved away to give us some space to discuss the matter.

Taking her gently by the arm, I asked her to walk with me. As we started to walk, I spoke with a relaxed voice.

"I know that you want to go with us. But . . . this could become a very dangerous life-threatening mission. I must tell you; it is difficult enough for me to think of any of our troops down, wounded, or dead if things do get nasty out there. We were very fortunate that the last two times no one was hurt. But, the thought of you or any other woman being injured or killed out there is one I don't care to deal with."

"Look, Lieutenant, I understand and I appreciate your concern for my safety," she said quickly, "but this is my job. I've been all over this country, been in combat situations, and I'm not afraid!"

"I believe that," I said. "But I am. I'm afraid of how I might act or react with you out there. I would be wondering where you were and how you were. And, if something happens to you, I will feel responsible."

"But you aren't responsible for me. I am responsible for myself, and I came to Nha Trang just to cover this story. It's a good story and there aren't many of those. Just let me do my job."

It seemed obvious she was prepared to argue the matter all day, if necessary, to get herself on what would become the press chopper. In my mind, though, that simply wasn't going to happen.

"Well, I can tell right now that we could go back and forth about this, probably until dark," I said. "But we've got to get this mission started now. I have my job to do." Then, I turned and started back toward the choppers.

She was following me back as we walked, so I talked over my shoulder.

"Look, it may be because of the way I was raised. I grew up in the South and was taught that ladies are special and to be treated with respect. I don't know; I just can't take a lady into a combat situation. I'm very sorry, but you'll have to cover the story from here."

When we reached the chopper, she marched around in front of me and positioned herself firmly.

"Damn it, Lieutenant! I'm no southern belle and I'm certainly no Scarlet O'Hara!" she snapped.

I have never heard the word "Lieutenant" spoken in such a way that it sounded like a very dirty word. Anyway, I turned to Orians and the others, "Okay, let's go. You all get on that one," I said, pointing to the second chopper from the front. "Once we secure the LZ, they'll bring you in."

Then I signaled down the line that we were ready to go and everyone began to load.

With the choppers powering up for lift-off, I turned back to check on the female reporter. She was standing with her arms tightly folded and didn't appear very happy.

"Look," I said. "Why don't you go up and begin visiting with the Montagnards? You can get a head start on your story."

Just then, Thieu Ta Ngoc walked up.

"Having trouble with a woman, Trung Uy?" he asked.

"I don't know, Thieu Ta. I suppose I am. Then, I thought— *This is certainly trouble I never expected to encounter in Vietnam.*

He laughed as we turned and walked toward the lead chopper. As we walked, Thieu Ta quipped, "You know, Trung Uy, we have this problem in Vietnam also. You are a good adviser. Can you advise me about this subject?" Clearly, he was amused by my dilemma.

"No, I don't think so, Thieu Ta. It doesn't look like I'm doing such a good job for myself."

When we reached the chopper, I was glad to see the familiar face of Captain John Wehr sitting in the pilot's seat.

"Ready to go?" he asked.

"We are. Get the Bandits fired up and let's go."

"Okay, but there's a change," he said. "Today, I want the entire company to be represented, so my callsign will be Intruder One."

"Good enough for me," I said. "Take us west Intruder One."

Just then, Ngoc walked up. "I'll be right here if you need me," he said.

I told him I hoped we wouldn't need him or the reinforcements and looked down the line of choppers to make sure we were ready. The ships were all at full power and everyone was loaded.

Turning back to Ngoc, I gave him a salute and told him I would see him later. He returned one of his waving salutes and said, "I will be here when you return this evening. Don't get any holes in your patch."

I laughed and jumped aboard Bandit 113, put on my headset, and said to Captain Wehr, now Intruder One, "Let's go!"

Even though having made this run twice before, I felt my heart rate increase as the choppers increased power and began to lift off. Adrenaline was already beginning to flow. As we started down the short runway, I glanced back toward the ground to locate the female reporter. I didn't see her along the runway and assumed she had taken my suggestion to begin visiting with the Montagnards.

Captain Wehr immediately took 113 into the lead of the wedge formation of our troop-carrying helicopters with Dolloff and the Wolf Pack close behind in their own mini-wedge of three. Once again, we headed west through the early morning sky. And, once again, Mang Quang was at my side. Even though I told him that he should stay at Trung Dung with his family, Mang insisted on going with me and—he wasn't going to have it any other way.

The day before, while making final plans, I walked to the Montagnard shelter looking for Mang Quang. He had done enough, more than enough. This mission could be even more dangerous than the last two and I didn't want him to be killed after what he had done for all the others. When I found him, I asked Ahat to tell him that I would be taking one of the other Montagnards as my guide on our final mission. I told Ahat to tell Mang that he had done enough and that, now, he needed to stay with his family.

Ahat had barely finished translating when an emphatic motorboat, "Na! na! na! na! na!" sound rippled from Mang Quang's mouth. Then, he quickly said something to Ahat that I couldn't understand.

Ahat replied to him in English, "I go too."

Immediately, Mang turned to face me squarely and insisted in English, "I go too, Trung uy!" He had added the word Trung uy, having learned that was the way the Vietnamese addressed me. Then, he showed me a huge smile.

I didn't argue with him. If he wanted to go, he could go—he had earned the right. And, I would be very glad to have him with me. However, I had Ahat tell him that he would ride a different chopper than mine this time. Because I was going in first and because this was our third trip, I didn't want anything happening to him in case the enemy was waiting for us this time. He would come in after the LZ was secure. Reluctantly, he agreed to that as a compromise.

On Our Way into the Mist and Fog

Our mini armada was once again on its way west. It was going to be a very long day, especially for me and a few others.

Entering the Valley of the Tigers

Just before reaching the LZ, Wehr directed the press chopper and the troop-carrying choppers to orbit and wait for instructions while we continued ahead with the Wolf Pack.

On this trip, rather than approaching from the north as we had done previously, we were going to make a low approach from the east. We would make our first pass perpendicular to the ridgeline, in case 18B's 7th Battalion was down there or if the enemy soldiers left in charge of the village had organized resistance. If either scenario were true, we expected our flight might draw some ground fire.

If we received more ground fire than the Wolf Pack could suppress, the F-4s in Cam Ranh Bay and "Spooky" were ready to respond and could reach us in minutes. Because of the extreme danger in today's mission, I had arranged for a dedicated commitment. Which meant they were on standby for our mission alone, they would be available whenever needed.

I had used F4's to get out of a jam before and Spooky was another favorite "equalizer," often used by our team. As mentioned, Spooky was the customized Air Force C-47 gunship that could pour out thousands of rounds of red streams of hot copper and lead, one round in every square foot the size of a football field. The deep droning roar of Spooky's miniguns was a welcomed sound for friendly troops on the ground. However, for enemy soldiers, it was often the last sound they heard.

While powerful air support was available, any battle would make our effort to recover the last of the Montagnards considerably more challenging. As we approach our LZ, I prayed our luck would last—just one more day.

As we passed over the ridge just east of the LZ, our ship was at the point of a diamond-shaped formation. There were two Wolf

Pack gunships just to our rear, staggered left and right, and one directly behind us, but farther back than the other two. Wehr led our small formation down through thin clouds and jungle mist to treetop level as we approached the cornfield. His approach was bold and confident.

The jungle was whizzing by just feet beneath us and adrenaline must have been pouring into my system because I couldn't wait to get on the ground, an absurd thought when you stop to think about it. There were people down there more than willing to kill us.

Wooosh! Like a great bird of prey, we swooped directly over the LZ as we made an east-to-west pass across its southern edge. We were going in on our original LZ and, from my position in the left door, I had an excellent view of the entire cornfield. Quickly scanning the tree line that surrounded the cornfield, I watched for muzzle flashes or enemy troops. Surprisingly, I saw neither. *Could we be this lucky?* I wondered.

"I didn't see or hear anything. Did anyone else?" I asked into the mic on my headset.

Wehr and each of the crew responded in turn, "No."

Then, Wehr asked the same of the flight and all the responses came back, "Negative."

Hearing that, I said, "Okay, call the slicks and put us in."

Wehr quickly radioed the slicks and flew a large circle.

By the time we were again lined up on the LZ, the slicks were right behind us—we were headed in one last time. And, today, mine would be the first boot to hit the ground and the last one off when we were extracted.

Captains Dolloff and Black had done this twice before. So, the gunships, as if practiced and synchronized, swung around behind us and changed formation to provide cover. If you had to be out in the middle of nowhere, Dolloff, Black, and the other pilots and crews of the Wolf Pack gunships were, without question, the men to provide your protection. They stayed right with us, then circled overhead as we settled in on the LZ.

Just before pulling my radio headset off, Wehr passed a final message, "Be careful!"

"I will. Thank you!"

Today, after we were inserted, Wehr would return to Trung Dung to pick up a second unit. They would be inserted on the lower LZ used by me the last trip, then Wehr and the Bandits would standby at Trung Dung.

I yanked my headset off and jumped to the ground with Ahat right behind me, as he always was. Following was Trung si Nguyen and six of Thieu ta's best Vietnamese SF troops he assigned to be with me during the mission. As they were jumping out, I moved to the front of the chopper where I could be seen. From there, I gave Wehr a thumbs-up, indicating he was clear to leave. He lifted off as the small squad fanned out to various points around the LZ. Immediately, the other slicks began arriving.

The chopper with the media people arrived first, and the pilot checked in to see if we were ready for them.

"Zero Two, this is Bandit Two-Five. I've got your press folks on board. Are you ready for them? Over."

At that point, there were only nine of us on the ground, so the LZ wasn't nearly secure enough.

"No, I'm not. Circle back and get at the end of the line, then bring them in. Over."

"Roger. We'll be back."

As he responded, he rolled his chopper out of the incoming formation so there would be no delay for the one following him.

Troops from the incoming choppers quickly helped secure the LZ, just as they had on the two earlier missions. Then, after the last troop-carrying chopper with Mang Quang aboard dropped him off, the one with the media aboard checked in once again.

"Zero Two, this Bandit Two-Five on approach with the press. Are you ready? Over."

"Roger, Bandit Two-Five. Bring 'em in."

After marking a set-down location for them, I pulled my collar up and turned my back to shield myself from debris churned up and blown around by the spinning chopper blades.

When the chopper lifted off, I turned back to make sure everyone had gotten off safely.

They had—everyone! Standing directly in the middle of my LZ was the female reporter who I believed was safe at Trung Dung writing her story. I was only slightly surprised when I saw her standing in the whirl of dust, grass, and cornstalk pieces stirred into a cloud by the departing chopper.

During take-off at Trung Dung, when she wasn't anywhere to be seen along the runway, I believed she had taken my suggestion and gone to visit the Montagnards. But, on the way out, considering her feistiness, another possibility came to mind—*What if she jumped on the chopper anyway?*

Well, that was exactly what she had done because—there she stood.

While still concerned about her safety, I wasn't truly disturbed when she appeared. She had convinced me that she was probably very capable of taking care of herself. But, because she was by herself, I was concerned that we might lose track of her if something unexpected happened. So, as the press chopper lifted off, I went over to speak with her.

Even when I got close enough for her to hear me over the roar of the departing chopper's engine and the loud beating sound of its blades, I had to speak loudly. Squatting down beside her, I shouted a question.

"Lose your way to the Montagnards?"

Smiling as she shouted her response, "Yes, I suppose I did. Are you going to shoot me for disobeying?"

"No, but the NVA or VC might. You know I could easily send you back on that chopper," pointing toward the one she had just gotten off of and now headed back to camp.

"But you won't, will you?" she asked.

"No, you can stay," I said.

As I spoke to her, the wash of hot air from the departing chopper blew her short hair back and forth across her face. I watched her as she brushed strands of hair away from her eyes. Looking at the soft features of her face, I was momentarily distracted by emotions that were ill-timed for this place. Regaining focus, I finished my direction to her.

"But, now that you are here, don't let your presence be a danger to any of these men."

She nodded that she understood as I continued.

"Stay up here on the cornfield, don't get anywhere near the jungle. If anything happens, we may not have much time to look for you."

Again, she nodded.

"If anything does happen," I told her, "find me or one of the other Americans and stay with him."

"Okay, I will. I promise. And . . . I'll stay up here. I just wanted to be out here so I could cover the rescue action firsthand. Truly, I won't get in the way. You won't even know I'm here."

"No, I'll know you're here. Just take care of yourself."

"I will, and thank you for letting me stay."

"Yeah, right."

As Trung Si and American advisers organized our command post, I walked over to the edge of the cornfield to a point where the hills and valleys to the west and north could be observed.

If our delay in returning has convinced the VC we weren't coming back, I thought, *it may have convinced the Montagnards as well.*

I was concerned that those for whom we had come may have decided to seek safety even deeper in more remote areas somewhere out there in front of me. If that were the case, I knew there would be little likelihood of finding them.

I called out to Sergeant Sotello, our medic who was running wires to a large set of speakers we brought with us, "Doc, when you get those hooked up, bring them over here . . ." Then, pointing toward the northwest, I continued, ". . . and aim them out that way first."

The valley floor ran in that direction, and it was probably the easiest path for families trying to flee from their captors.

While Tracer would fly over that area, we were going to support his effort by using Mang Quang to reach the still missing villagers. He would broadcast a message live from the LZ to ring out over the valley. We would be ringing the bell of freedom one last time and we would be ringing it loudly. I hoped the missing Montagnards would hear it or were close enough to hear it.

Off and on during the day, I had our Vietnamese interpreters take either the microphone or bullhorn to invite the VC and NVA to surrender. I didn't truly expect any of them to accept, but I thought I would give them the opportunity.

From my vantage point on the ridge, I could see the second

group of troops being inserted by Wehr and the Bandits on our second LZ. If there were going to be a battle, I would have the superior force. Without a clear view of the LZ itself, I couldn't tell if they had started up the ridge. Just then, Light, an alternate radioman and interpreter, walked up and pointed down toward the other LZ.

"Trung Uy, those guys just called. They said their point unit is moving into the jungle."

"Good, let's go help Doc get the speakers going."

It took only a few minutes to finish hookups and position the speakers, but it took much longer for them to yield results. Unlike the quick response we had experienced during the previous two trips, after an hour or so of broadcasting, we still hadn't seen the first villager.

Squad-size patrols were positioned around the LZ for security while others searched for villagers and the enemy. During that time, the colonel from 5th Group Headquarters, Frank Orians, and the media had little to do but wander around on the cornfield.

Something very strange occurred to me as I scanned a group of people walking around on the LZ. I realized there were people in American uniforms that—I didn't recognize. I had no idea who they were or where they had come from.

Curious, I walked over to the group, introduced myself, and asked, "Who are you men?"

As it turned out, they were young low-level staff members from 5th Group Headquarters (clerk typists, drivers, and communications specialists) and they had come out with the colonel. Seems talk of the rescue had gotten around Group

Headquarters and several of the men talked the colonel into letting them come with him. One of the men spoke up and said, "When we heard this was a rescue mission, we wanted to help. We never get to do anything meaningful at headquarters."

What was I going to say? I certainly understood them wanting to do something meaningful. So, since they were all in combat gear, probably for the only time during their Vietnam tour of duty, I said, "Okay you guys are now part of our LZ security team. Spread yourselves out around the perimeter of the LZ and watch the jungle, but don't go into the jungle. And, DO NOT shoot anyone who is not shooting at you. We expect villagers to be walking in from anywhere around the LZ and I don't want you to shoot any of the people we are here to rescue."

The Colonel and 5th Group Troops

Here, I am showing the Colonel's men where to position themselves around our perimeter. They were ready and willing to join the mission. I was amused and they seemed very excited to become an "official"

part of the mission. At a minimum, they would have a story to write home about.

And, as you can see, Mang Quang was never far away from me.

David Culhane and his crew were filming background footage of the valley and security troops around the edge of the cornfield. Don Tate, the female reporter, and the others were all making notes and snapping pictures of our surroundings.

I watched the lone female correspondent as she confidently walked the cornfield in search of her story. She didn't know it, but I had assigned an American team member to watch her and stay near her. Occasionally, she would stop and shade her eyes as she scanned the surrounding jungle for emerging villagers that were yet to be seen. One time, when she was very near the jungle edge, she turned as if to get her bearings. When she saw me watching her, she waved me off as if to say, "Quit watching me, I'm okay."

But she wasn't okay. She was unarmed, too close to the jungle, and too far away—even though I had someone shadowing her. So, I waved her back. Clearly remembering her promise to stay near the center of the cornfield, she shook her head in submission and started back.

After more time passed, I was beginning to be concerned that the media representatives wouldn't have a story to photograph, write about, or tell beyond the little bit they had already seen and heard. Frank Orians had considered the same possibility because he walked over with Culhane and some of the others.

"Lieutenant Ross, since not a lot is happening, can you give these folks some background information on the operation?" Frank asked.

Recognizing that it wasn't a question, but a request to give them something to report, I said, "Absolutely. What can I tell you?"

The very first question was quickly posed by David Culhane who asked, "What is the name of this area, Lieutenant? Where can I say we are?"

An intelligent answer to that question was going to require some quick thinking.

CHAPTER 18

In the Wild

JUST BEFORE LEAVING FORT BRAGG for Vietnam, I was assigned as an operations officer and referee for a large joint unit training exercise. Headquarters for the exercise was set up in a field not far from a small airstrip in rural North Carolina.

Early one morning after we had everything set up and were waiting for the exercise to begin, I walked down to the airstrip and met the operator of a one-owner civilian flight school. I think his name was Bob and he was very cordial. After I explained what we were doing there and told him about the exercise, he told me about his school and showed me the two planes he used for training, both were Piper Tri-Pacers.

Then after I told Bob that my Dad served with the 8th Air Force during World War II and that I had always wanted to fly, Bob quickly turned into a salesman, "I can teach you to fly if you still want to learn," he said.

I laughed and told him that I didn't think I would have time, but as it turned out, I did. For a couple of hours after sunrise, while training units maneuvered and prepared for the day's exercises, things were very quiet. So, with nothing to do but wait for the day's exercise to begin, I walked down to Bob's office and signed up for flying lessons. When he asked when I wanted to start, I said, "What about right now?"

He laughed and said, "Let's go!"

Then, sitting in the seats of one of his Tri-Pacers, Bob ran through flight basics and discussed a list of safety issues. After that, we took to the clear North Carolina blue sky. That was my first time in a small plane and I loved it.

Over the next several days, sometimes twice a day, I would visit Bob whenever I had an hour to fly. At the end of my seventh hour, we landed and Bob asked me to taxi back to the end of the runway. At the turn-around, he then asked if I had the learner's license he had given me when I signed up for lessons. When I said that I did, he asked for it, took a pen out of his pocket, and signed it. Then, he said, "You're ready to solo," and got out of the plane!

Just before he closed the door, he said, "Have a good flight. Don't go too far away. I want you to be able to find your way back with my airplane."

For a moment, I sat there amazed and imagined that Bob must have very good insurance and maybe he wanted a new airplane. But, I quickly shoved the throttle forward and headed down the runway. Just before I reached the end, I pulled back on the yoke (steering wheel) and lifted into the sky—alone. It was an incredible feeling.

For the next hour, I flew large circles around the airstrip, always keeping it in sight. I flew high and I flew low. I soared and dove with the freedom of a bird and didn't want to land, but the day's training exercise would soon begin, so I headed back to the airstrip.

During staff training briefings later in the day, I learned that my friend, Lieutenant Bill Phalen, and his A-team would be camped near a train trestle the next morning. Their training mission was to destroy the trestle to prevent imaginary enemy supplies from reaching field units. Of course, there was no real enemy supply train and they wouldn't actually blow up the railroad trestle but would mark it with red tags to indicate that it had been destroyed for training purposes.

Knowing where he would be, I thought I would put my new flying skills to practical use the next morning, try to find the trestle, and see if I could locate Phalen and his team.

Since the trestle was several miles away, I marked a path from the airstrip to the trestle on an exercise map. I also marked the roads from the airstrip to the train tracks that led to the trestle. I thought that, if I got disoriented, I had more than one way back to the airstrip.

Shortly after sunrise the next morning, I took off on my second hour of solo flight training and my first "unofficial" military observation mission. I used the roads as my guides until I could see the train tracks, then simply traced them until I could see the trestle. Once over the trestle, I circled, then dropped altitude and crisscrossed the area looking for Phalen and his team. They were nowhere to be found, at least from the air. If they were down there, they were doing their jobs as Special Forces operators and were staying hidden.

After about thirty minutes of searching, I gained altitude and flew a reverse path back to the airstrip. That was my last day of flight at Bob's school. The exercise ended late the next day and I returned to Fort Bragg, but it wouldn't be the last time I piloted a small fixed-wing plane and I would fly it out into enemy territory.

My job at A-502 required me to know our AO physically as well as what was happening within its borders. To meet those important requirements, I flew many observation missions over it.

One of my early missions was with an Army 0-1 (small single-engine plane) pilot, he sat in the front seat and I sat behind him. As I was strapping myself into my seat, I noticed that there was a steering stick clipped to the side of the plane next to my seat and a slot in a box on the floor between my feet that the stick fit into.

After we took off and were gaining altitude, the pilot asked, "Where do you want to go, Lieutenant?"

"If you let me put this stick in the slot where it goes, I'll take you there," I said.

He laughed and said, "Do it. Let me know when you're ready and I will release control up here."

We were headed north off on the Trung Dung runway, but after the pilot released his control, I turned west and we headed out toward My Loc. Intelligence information indicated that there was new enemy activity beyond our westernmost outpost and I wanted to see what we could find.

After we flew over My Loc and farther out, I told the pilot that we were looking for trails or any other signs of enemy activity in the area where we now were. I said, "We probably need be as low as we can get, so I will let you take over and do what you do."

"Roger, I'll take it," he said, ". . . we're headed down."

We flew low over and around hills as well as along and down in river bottoms. We had been at it for a while and were flying at treetop level over a river when I noticed a tiger jump away from the river's edge when it detected our approach. The big cat was beautiful and looked as if it had just been groomed. The white on its belly was snow-white and the orange on its back was vibrant. The stripes that wrapped the animal were jet black and they seemed to shine. This was the first tiger I had ever seen in the wild and it quickly disappeared into the jungle that surrounded the river. Eventually, I would see three of them.

I saw the second one while flying an observation mission with the 281st out in the same general area while on another hunt for the elusive enemy. This one was also walking the river's edge when we happened upon it. We were following Song Cai River west and were flying low, down in the riverbed below the treetops, and turned quickly to follow a curve in the river. When the whop, whop, whop sound of the chopper's pounding blades suddenly echoed through the riverbed, the tiger leaped for the cover and safety of the jungle. In little more than two bounds, the magnificent animal was gone.

The third one, I saw during another mission with the 0-1 pilot. I'm sorry I don't remember his name because I enjoyed flying with him. When he arrived to pick me up that day, he had already installed the control stick in front of my seat. When I got in, he said, "I know you like to fly and thought you could take us wherever we're going today."

I laughed and accepted his offer. As soon as we cleared the end of the runway, I took control and once again headed west. This time, we flew far beyond where we had flown before. My thought was that the farther we flew from allied positions, the more reckless the enemy might become and we could get lucky and catch them in the open.

We were flying through a valley that was a hundred shades of green and looked like it could be part of the Hawaiian Islands when we saw a tiger crossing the open valley floor. Again, the big cat was beautiful and it stood out against the green of the lush jungle that surrounded it. Seemingly undisturbed by our presence, it only glanced up at us, then continued on its way as we flew over it and continued through the valley toward the Cambodian border.

Each time I saw a tiger in the wild it was a special experience and it was wonderful to see them running free. Before seeing these, I had always associated tigers with animals you only saw in a zoo.

In only a few months, I would be responsible for naming the beautiful valley we had just flown through and the name would be broadcast internationally.

Memories of that valley and seeing the tigers not so far away from the Montagnard village would give me an answer to Culhane's question about our location.

View from the Back Seat

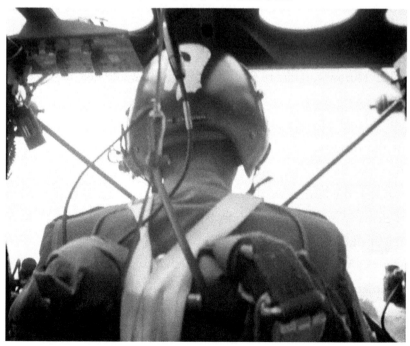

This was my forward view from the rear seat of the O-1. However, I had a clear view of the terrain out both of my side windows.

CHAPTER 19

Middle of Nowhere

MY FIRST RESPONSE TO Culhane's question, "Where can I say we are?" was an honest one. But after I quipped, "In the middle of nowhere," I pulled the new map out of my web gear. "Until yesterday," I said, "I didn't have a map of this area. On the previous two missions, we had flown off our usual map and depended on our guide."

I looked for any location name on my map, but there were none. *Do I make up a name?* I wondered. Then, thinking for just a moment, I responded with another honest answer, "Well, we've flown so far west of the populated areas that there is no specific name for this place. However, on previous observation missions out this way, I've seen tigers on the prowl either along the river bank or crossing open fields near the river. So, I simply call this entire area . . . The Valley of the Tigers." I loved the name and so did Mr. Culhane.

"Great name!" he said, "That's where I'll say we are."

When I later had the opportunity to watch David Culhane's CBS report on the rescue mission, it made me smile when I heard Walter Cronkite or one of the other newscasters describe the rescue as having taken place in "The Valley of the Tigers." And, it was also amusing to discover many years later that CBS had cross-filed the raw footage of Culhane's rescue report in their archives under "Rescue in the Valley of the Tigers."

The questions after "Where are we?" were easy. The reporters wanted to know how the rescue effort had been initiated and asked why we were here today. Since things were quiet, we discussed much of what had occurred over the past several days. During our visit, I suggested that they talk to some of the other advisers for additional background information. They were also told that, if and when villagers appeared, they could use any of our Montagnard interpreters to obtain firsthand accounts for their reports

There was virtually nothing newsworthy left to discuss when things finally started to happen. As our mini-press conference was breaking up, Light walked up with a message from one of our patrols. The small unit surprised four or five VC sitting along a trail and had captured them and their weapons without a shot being fired. They were bringing them in.

Early in the first mission, everyone, Vietnamese and American, had been cautioned about shooting too quickly. We knew the Montagnards might appear almost anywhere, and many were dressed in khaki or black pajama–type clothing, very much like that worn by the NVA and VC.

The patrol making the capture had done their job under

extremely difficult circumstances and with great restraint. Once they identified the individuals on the trail as VC, they could have easily opened fire. Rather, they chose the more difficult and dangerous option of capture and had been successful.

Shortly after receiving news of the VC capture, one of the Vietnamese soldiers on the perimeter alerted us to a small group of Montagnards that was approaching from the lower part of the ridge. All of the media people immediately ran for cameras and equipment and headed over the hill to begin documenting the story that finally seemed to be developing.

At almost the same time as the first villagers emerged from the jungle, one of the patrols radioed with a report that they had encountered several others trying to make their way to the LZ.

As villagers popped onto the cornfield from various points around our perimeter, the CBS team moved around on the LZ, filming as missing families made their way up the hill. Other members of the media were either taking notes or pictures as they watched the arrival of Montagnards who carried belongings in baskets on their heads, backs, or under their arms. Men, women, and children were hurrying onto the cornfield.

There was new fervor in the voice of Mang Quang who, via the loudspeaker for the past few hours, had been encouraging his friends and other families to come to the LZ. Then, just as his effort was renewed, the speaker went dead. Something had happened to it: either the batteries had become weak or it had overheated from use, but that didn't stop Mang Quang. He grabbed a bullhorn we brought with us and continued to encourage his people to try to reach the cornfield.

His determination to help his people had paid off. One of the next people to emerge from the jungle was the brother of one of our Montagnard interpreters. They hadn't seen each other in years.

The opportunity to witness their reunion was a heartwarming experience for all to see. Of course, the media loved it. And, when other villagers arrived with smiles on their faces, even the colonel from headquarters, a seasoned military man, walked over to say how glad he was he made the trip. "Seeing what I have seen has made my trip and the wait worthwhile," he said.

Villagers continued to arrive on the cornfield while our patrols found others in small groups of twos, threes, and fours. This continued until just before noon when I asked for a count. Three more had just arrived, and according to my tally, we had forty-two people, which seemed to account for everyone on Sergeant Koch's list of the missing. Based on our final discussions with the Montagnards, we were expecting to find between forty and fifty more villagers. So, when the count totaled forty-two, everyone was very pleased.

"Great!" I said. "That should be just about everybody. We'll wait a bit longer, but let's prepare to move some of these folks out of here. We'll start transporting villagers back to Trung Dung now."

We still had a unit on the ridge west of us, and several security patrols were still roaming the area. The rain was approaching from the east, so we needed to start moving, not just villagers, but the media, and troops as well before the rain closed in. If there were any straggling villagers still in the jungle, they would have plenty of time to reach the LZ.

We had been very lucky up to that point. While there had been minor encounters and brief firefights with enemy units during the day, we had been able to recover all or most of the missing villagers. At the same time, seventeen VC had either surrendered or been captured without serious injury to either them or us. There was no need to press our luck this late in the mission, so I had all of our units recalled.

Next, I called Intruder One, Wehr, and requested the slicks for extraction. We would start moving the Montagnards while we waited for our other unit and patrols to arrive.

Culhane was standing nearby and heard the call.

"Lieutenant, I'd like to interview you on camera before we leave. Is that all right?" he asked.

"Sure, let me know when you're ready."

"How about right now?" he asked.

"Fine, let's do it."

He gave his crew some quick directions, and in just a few minutes he was ready. "Lieutenant, we're set. Would you stand right here, please?"

Culhane had barely finished his interview when Wehr called to let me know that he and the slicks weren't far away.

With the missing Montagnards seemingly accounted for and about to be picked up, Tracer's work was again finished. Looking out over the valley, I could see him running the valley bottom, and he was still broadcasting.

"Tracer, Zero Two. Over."

"Roger, this is Tracer. Go ahead, Zero Two."

"We've got 'em all, Tracer. You're clear to head home."

"Roger, understand. I'm coming up." Tracer pulled up sharply out of the valley bottom and turned toward the cornfield. As his plane got closer, I pointed toward it and told some of the media people who were standing nearby about his role in the rescue.

Then, just as he zipped overhead, I said, "That guy is one of the people who made this mission a success."

Tracer Passing Overhead

As he passed over the cornfield, Tracer still had his speakers at full volume. So, all of the media could clearly hear Mang Quang's message being broadcast.

"What's his name?" Don Tate asked.

"I don't know, but we can find out easily enough," I said.

"Tracer, this is Zero Two. Before you disappear into the wild blue yonder, we'd like to know your name, over."

"I'm Major Ken Moses," he responded.

"Ask him where he's from," the female reporter said.

"One of the reporters wants to know where you're from."

"Rush City, Minnesota," he responded proudly. Then, he added, "If you don't object, I'll stay in the area just in case you need me as a FAC."

"Good idea," I said, "You are welcome to stay as long as you like."

"Great! I'd like to see you get everyone out and on their way."

Since that day, I've thought about the man with the callsign "Tracer" many times and what he did. But I never met him face-to-face and never spoke to him again after that last day of the rescue.

While I was in Vietnam, there many very brief encounters with people I still have not forgotten. Many of them I have written about.

I was still watching Tracer fly a wide circle overhead when one of our patrols called in. They reported seeing a VC unit of approximately twenty-seven men but also reported seeing regular NVA troops with them. That information troubled me. I was concerned that elements of 18B's 7th Battalion might be arriving in the area as expected. Dealing with the NVA would be much more challenging than the VC we had thus far encountered.

Just before leaving camp, Ngoc had reminded me that they were due any day. We didn't need a fight at that point; villagers and members of the media were all over the cornfield. The mission was essentially complete, and extraction choppers were due at almost any minute.

Because the report had come from our Vietnamese patrol, I had one of the other advisers call the Americans with the unit on the ridge west of us to alert them to the situation.

That the VC/NVA unit had been spotted moving up the gully between us didn't disturb me. If there was going to be a fight, they made a serious mistake. They had moved to a position between our two units. We could block the gully from above and below with patrols we already had out and simply close in on them from both flanks (sides) if that became necessary.

What did disturb me was the fact that since they knew the area well, it was unlikely that they would make such a mistake and may simply have been trying to distract us while a supporting unit maneuvered around us. Under the cover of the thick jungle growth, they could be attempting almost anything.

To determine exactly what they were up to, I quickly redirected one of our small patrols in an attempt to again locate the unit.

Then, anxious to move the Montagnards and press folks out of harm's way and back to Trung Dung, I called Captain Wehr to alert him to our new situation and asked about the slicks. "Intruder One, Zero Two. Over."

"Roger, go."

"Intruder One, what's your ETA? Over."

"Look to the northeast. We're on approach now. Over."

"Roger, I see you. Perfect. And, Leader . . . keep the guns in close. Charlie is nearby. Over."

"Roger. Understand. I've been monitoring and so have they. We'll keep you covered. Over."

"Roger, thank you. Out."

With slicks on approach, I asked Sergeant Trujillo, one of the Beast's creators, and among our American advisers, to mark a touchdown location for the arriving choppers. Then, the media people were told to get ready to leave with the villagers.

During the time we were looking for and waiting for Montagnards, then when we were preparing to begin leaving the area, Captain Dolloff and the Wolf Pack were patrolling the entire area. They were flying a pattern that allowed them to cover both ridges where we had troops.

Dolloff and his crew as well as the pilots and crews of the other two Wolf Pack gunships could monitor the transmissions between our ground units. So, whenever the ground units reported enemy activity, the Wolf Pack shifted their flight pattern to be closer to the reporting unit in case they were needed.

While everyone else was gathering equipment, Culhane walked over to me. "Lieutenant, I'd like to stay a little longer, if that's possible," he said.

"I don't think that's a good idea. We know that there is at least one enemy unit moving around at the base of the ridge. You may get more of a story than you bargained for, but if you want to stay a few minutes longer, that will be your call. Just make sure you are aboard the last chopper out. You won't want to be here after that."

"Fine, I will . . . and I'll keep my crew out of the way."

He told his crew they were staying until the last chopper arrived and directed the cameraman to start filming the slicks, which were about to touch down in the area that Sergeant Trujillo was marking for them.

Then, just as the slicks began arriving, the possibility of a disaster suddenly developed. The small patrol I had redirected to find the NVA located them moving toward the LZ. I quickly called Captain Wehr and told him we needed the Wolf Pack and we needed them—now!

Obviously hearing my transmission, Dolloff immediately organized the Wolf Pack.

Our patrol marked its location and gave me the position of the enemy unit. Then, the Wolf Pack took charge and did, in fact, once again deliver hell from above. One after another, the three gunships poured rockets and minigun fire into the enemy position and all around it. White smoke billowed up out of the dark green jungle.

Culhane immediately directed his cameraman to film the action.

While Dolloff, Black, and the Wolf Pack worked on the enemy, I encouraged everyone to help get the villagers aboard the slicks as quickly as possible. We didn't need a tragic occurrence during the last hours of our nearly completed mission.

When families crowded around the chopper doors, team members, crew chiefs, and door gunners helped lift women and children aboard. But even as they helped the Montagnards onto the slicks, the crew members kept glancing up to keep an eye on the jungle around the chopper.

Boarding During a Battle

If you look beyond the tail of this Bandit helicopter, you can see the white smoke rising from Wolf Pack's attack on nearby enemy positions.

While the slicks continued to arrive and depart, the Wolf Pack pressed their attack on the nearby enemy position. The sky seemed to be packed with spinning airships. Some were ferrying the villagers out while others provided an umbrella of protection over those of us on the ground. *God, don't let them run into each other,* I thought.

Under the cover provided by Wolf Pack, the Bandit pilots and crews worked so quickly and efficiently that the Montagnards, the media representatives, the 5th Group colonel, other straphangers (those not officially assigned), and some of our troops were gone in just a matter of minutes. As the last of the Montagnards were ferried back to Trung Dung to be reunited with the others, Major Moses flew above and behind them—as if a shepherd. That would be the last time I saw him.

I was relieved to see the cornfield relatively clear of people and particularly pleased that the female correspondent was also on her way back to Trung Dung. One of the last times I saw her, she was moving through the huddled Montagnards writing on her notepad. Seemingly undaunted by our location in the middle of nowhere, she moved from one family grouping to the next, snapping an occasional picture and making notes. Even though unarmed, I never saw any sign of fear—only dedication and confidence in her ability to do her job.

At one point, watching her work, she reminded me of the woman taking pictures of combat action in Nha Trang the first day I arrived. They were both very self-assured. And, like Tracer, who had flown above the village every day we were there, that day would be the last time I saw her.

In Tribute to the Wolf Pack and all the Intruders

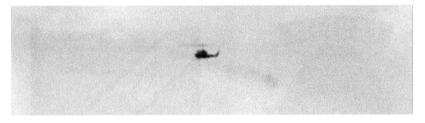

Above, part of the blistering attack. Below, part of the results.

The wolves howled! And, they watched over us and they shielded us from the enemy.

The 281 Assault Helicopter Company

It was an honor for me to fly with the courageous men of this unique aviation unit. They took me and the other Special Forces soldiers wherever we needed to go, then protected us and came for us when the job was done.

The Intruders

The men of the 281st were both incredibly daring and amazingly skilled. As noted earlier, the 281st flew the dangerous Delta Project missions of our 5th Special Forces Group. Flying in their UH-1 "Huey" helicopters, they were exposed to the world and enemy gunfire with very little protection. These men were, and remain, very special to me.

As you continue to read, it will become obvious why I feel about the 281st as I do. They were selfless men who wouldn't hesitate to risk their own lives for others—and that included mine.

Exceptional military units are typically led by very dedicated leaders, men, or women who ensure that those under their command are well trained and well cared for. John (Jack) Mayhew was the commander of the 281st AHC while I was in Vietnam. Then a major, Jack was such a leader and a man who cared deeply about the men who served under him and he still does. To them, he was the "Old Man." And, even now, more than 50 years later, I am told that Colonel Mayhew is still affectionately thought of as— "The Old Man-Intruder 06."

You can read more about Colonel Jack Mayhew and the 281st in his book, "BEEP! BEEP!"

CHAPTER 20

A Promise Fulfilled

WITH THE VILLAGERS ON their way to safety, my attention quickly turned to the security of our remaining troops and getting them safely back to Trung Dung. The Wolf Pack had ended their attack on the enemy position and everything was momentarily calm. It was our third day in enemy territory and I didn't want to lose anyone this late in the mission.

After Wolf Pack's display of power, there were no reports of either VC or NVA activity. I took advantage of the lull to consolidate our forces. The unit that had been on the ridge west of us was called and told to move to the cornfield. Smaller teams continued to patrol around our position since we needed both security and early warning. Those would be the last elements to be withdrawn from the jungle.

Our western unit was headed east from the ridge when its point unit discovered several huts. At the outer edge of the small complex, they encountered another small VC/NVA unit farther down in the gully. There was a brief exchange of light weapons fire and then quiet. The enemy unit had disengaged and quickly disappeared in the thick jungle growth.

My Vietnamese counterpart, Trung Si Nguyen, radioed his men and told them to burn the huts and any supplies to prevent them from being used by the VC or NVA. Nguyen also told his men to cross the gully and continue toward the LZ as soon as the village was burning.

Immediately upon igniting the village, white smoke filtered through the jungle treetops and, like the earlier rocket fire, billowed over the jungle canopy. As our unit continued to destroy the huts, the small enemy unit circled back and, while trying to salvage supplies, began sniping at our troops. With the huts burning, the unit was told to break contact and clear the ridge rather than continuing to engage the enemy. We had other plans for them. Hopefully, with our unit withdrawn, the enemy would continue their salvage efforts before the village burned to the ground. If they did—big mistake!

In the last picture, I immediately radioed for assistance from an equalizer that was very close at hand, much like a pine branch I once used many years earlier.

After our unit was given time to move away from the village and clear the area, I called Dolloff and asked that the Wolf Pack again do what they did well. I told him to use the smoke as a reference point from which to begin firing onto the village complex. Their fire would also provide cover for our unit as it made its way across the gully to the LZ.

Upon receiving my request, Dolloff and Black were in the led as the three Wolf Pack gunships immediately rolled around and, in a single file, began their attack. As before, one by one, they fired rockets and their miniguns into the mountainside. And, just as before, the rockets left thin lines of white smoke as they streaked into the treetops where they disappeared and then exploded. The miniguns created a deep humming sound as they rained a hellish stream of fire into the jungle canopy.

Wolf Pack's continuing attack successfully shielded our unit from the enemy, which oddly, seemed to have been in pursuit. It wasn't long before our unit's point element emerged from the jungle. They had crossed the gully successfully without any other enemy contacts.

However, as the unit continued to arrive on the cornfield, one of our circulating patrols at the base of the hill encountered another small unit they believed to be all NVA. But, again, there was only a momentary exchange of gunfire, and the enemy quickly dissolved into the dense jungle, with which they were intimately familiar.

Within minutes of the enemy contact report, the rear guard of the incoming unit cleared the jungle and almost simultaneously the slicks began arriving. So, as soon as they reached the LZ, our troops were quickly loaded and they headed back to Trung Dung.

With turnaround time at about forty-five minutes to an hour, it would be well into the afternoon before our extraction would be completed. With at least two VC/NVA units in the area, I decided to walk our entire perimeter, which was growing thin as troops were extracted.

While I was in the jungle, Lieutenant Sullivan, who was up on the LZ, called to share some news with me. He said that there were only two more choppers left in the current extraction rotation and they weren't far away. He said there wasn't going to be enough room for all of us, but Intruder One and other slicks had already headed for refueling. They expected to be back in an hour or less.

Sullivan's report wouldn't be my greatest concern. Considering that the NVA unit our patrol encountered might be part of the larger 18B unit we were expecting, I had asked to keep all the Wolf Pack ships on station with us. As a result, along with the Bandits, all three gunships now needed to refuel. So, when Captain Wehr and the Bandits left, Captain Dolloff and the Wolf Pack went with them.

I told him that I would keep Ahat, Trung si Nguyen, a couple of his key men, and Ngoc's VNSF soldiers. He could take whomever else he could get on the two incoming choppers and leave. "We'll be fine for an hour or so," I said. "The Wolf Pack did a good number on the bad guys and I haven't seen or heard a thing since I've been down here. And, they'll have no idea how many of us are still here."

"Okay, I'll load the choppers . . . but, I'm staying with you. Out."

Staying with a Teammate

Above: Mike Sullivan (R) and Vietnamese recon platoon member (L). By saying "Out" Sullivan wasn't waiting for my response, he wasn't leaving. He was going to stay with his teammate. The men of A-502 shared a very special bond and a similar bond was shared by members of virtually every unit in Vietnam. Your teammate could be the key to your survival and vice versa.

When all of the helicopters, including the Wolf Pack, and almost all of our troops were gone, the rugged mountainside became very quiet. After a while, only the call of jungle birds could be heard, along with the occasional roll of distant thunder.

After spending nearly an hour in the jungle positioning and being a moving part of our nearly nonexistent perimeter, I made my way back up toward the center of the LZ to wait for the slicks that were due to arrive at any time. Back on level ground and able to see across the LZ, I was surprised by what I saw.

There, near the center of the cornfield, stood David Culhane and his crew. When I left to go down over the hill, they were packing their gear. I was certain they had been taken out on the last lift. After getting close enough to be heard, I asked, "Why are you guys still here?"

"I told you we wanted to cover this operation as close to the end as possible," Culhane said.

"Yeah, well, this isn't close to the end. This is the *very* end. This was never a safe place for you to be, but it is certainly a very dangerous place for you to be right now. It isn't safe for any of us. If there had been two more choppers, we would all be gone," I said.

"But the choppers are due back about now, aren't they?" he asked.

"Yes, they are . . . but, do you realize how few men are securing this LZ?" responding with a question of my own.

"I don't know, maybe a hundred?" he guessed.

Then, I revealed reality to him. "No, there's me, my radioman . . . this adviser," I said, pointing to Mike Sullivan, ". . . Trung Si Nguyen and about ten of his soldiers. And, well . . . the three of you. That's it."

It was clear from the change in his facial expression that he instantly understood the seriousness of our situation as I continued, "And, Trung Si's men are only very loosely deployed around our entire perimeter to provide security until we are extracted. That's where I've been . . . there are a lot of holes in our fence."

Culhane said nothing but exchanged glances with his crew. As the three contemplated what they had just heard, Ahat, who had been with me in the jungle, walked up and said, "Trung Uy, Intruder One wants you on the radio."

"Good." I said to the others, "We'll find out where they are."

Taking the handset from Ahat, I responded, "Roger, Intruder One, this is Zero Two. How far out are you?"

"A long way," the voice said.

It wasn't Captain Wehr, though. It was the radio room at Trung Dung, and it sounded like Jean Lavaud. The radio snapped and crackled as he spoke, so I wasn't sure who it was. Something was causing interference.

"Intruder One is still on the ground, Zero Two. The rain is pouring down here and in Nha Trang, he asked us to relay. He has been trying to reach you for some time. He has no visibility and they can't fly. Over."

"Roger, understand. Any idea how long before he can get back up? Over."

"No. The rain is pretty heavy here right now."

Turning toward the east, it was easy to see the gathering storm clouds. While still in the distance, they were thick, very dark, and appeared to be moving our way.

"You know these storms," the unidentified radioman continued. "It could blow right through. Intruder One says they are all refueled and ready to fly. He said he was just going to sit and watch it for now. Over."

"Roger, just advise him of or situation. Due to the extraction rotation, I only have one American adviser, my radioman, and ten Vietnamese soldiers with me . . ." Before I could finish, David Culhane raised his hand just so I didn't forget him, which I hadn't. So, I continued, "Oh, and three members of a CBS News crew. Over."

"Wow! Okay, I'll pass that along to him right now. Over."

"Roger, we'll be standing by. Zero Two, out."

Grounded by Weather

When the villagers and most of our troops were being dropped off, the weather was quickly moving in at Trung Dung. It wouldn't be long before it reached those of us still out on the cornfield. You can see how low and dark the clouds were. That last chopper had to fly back through the weather to reach the Nha Trang Air Base.

Giving the handset back to Ahat, I turned to Culhane, who had been listening to my conversation with Trung Dung. "I told you. You should have gone when you had the chance."

"Yeah, I guess you're right. What do you think? Any idea how long we'll have to wait?" he asked, not inappropriately distressed.

"Do you want a straight answer?" I asked.

"Yes, of course."

Pointing toward the east, I said, "See those clouds? They aren't our only concern. It's going to be getting dark soon. Whether or not we get out of here tonight depends on how quickly those clouds move."

"Sure, okay . . . I understand," he said.

Now he seemed more anxious. Frankly, I wasn't very excited about our situation either. We had become a very small unit, now with absolutely no support and a very long way from home. On top of that, we were in an area that was a very familiar landscape to the enemy. Despite our dilemma, until there was a reason for concern, there was no need for Culhane or his crew to worry unnecessarily. I did what I could to reassure them. Then, trying to lighten the moment, I told them we might have to give them some quick jungle training.

"You may get more of a story than you came for," I said.

There were a few weak smiles and the cameraman said, "No, the story we have is great. Just get us outta here with it."

I laughed and again assured them we would cope with the situation whether or not the helicopters made it back. Deep down, I knew that our situation was a tenuous one.

Time seemed to pass slowly as I stared into the approaching storm that would soon be on top of us. Privately, I shared my concern about our defensive positions with Sullivan. He said he would keep an eye on the perimeter and headed down over the hill.

With Culhane and his crew on the LZ, there was little for me to do but wait and watch the distant clouds roll closer and closer. I did what was possible to keep spirits elevated. Eventually, though, clouds filled the valley. It was difficult to imagine how the choppers would be able to fly through their dense accumulation.

As the wind began to blow and rain started to fall, an extraction before dark began to seem extremely unlikely. Since our enemy knew exactly where we were, and because extraction might not be possible, it seemed apparent that I needed to start thinking about moving our small unit off the LZ and into the jungle.

Pulling my map out, I looked down toward the valley bottom. Running my finger along the terrain features, a route was traced down to a clearing that was barely visible in the distance. If we weren't extracted before dark, we would spend the night in the jungle and move toward the clearing in the morning. It would serve as an LZ for our extraction the next day.

With a contingency plan prepared, I refolded the plasticized map Sergeant King had given me and slid it back into my web gear. There were times when I thought King was psychic, and this was one of them. His map was already proving useful. Looking up and out across the cloud-filled valley, I became aware of the light rain blowing in my face. *At least my map won't get wet,* I thought.

The rain was cold, which was a strange feeling. I couldn't remember feeling cold at any time during my many months in Vietnam.

Maybe it's cold because we're in the mountains, I thought.

Turning to Culhane, I asked, "Everything okay?"

"Wet, but okay," he replied. "Lieutenant, you know this film is valuable, and we need to get it out of here as soon as possible. This is a story I'll enjoy telling. There aren't many of those, you know."

It seemed increasingly unlikely to me, as cold rain dripped off my nose and chin, that any of us would be going anywhere by helicopter that day. But, until such was a certainty, being pessimistic wouldn't help our situation. So, my attitude continued to be cautiously optimistic when I responded to Culhane.

"I understand," I said. "And you shouldn't worry. I want to get your film and our hides out of here as quickly as possible. Believe me, I don't want to hang around here any longer than necessary. But, with the rain and clouds, it may take them a little longer to get back to us. Even if they could fly it would be difficult, if not impossible, for them to find us now."

"Well, I just hope we're not out here overnight," Culhane said.

It had been about thirty or forty minutes since we communicated with Trung Dung when my radio began to snap and crackle again. It sounded as if someone was trying to call, but we could hear only bits and pieces of the transmission. Ahat turned the volume on the radio to its highest setting. Problems began to compound; the radio battery was getting weak and we had already used our last spare.

Mixed with interference, we could hear, "—ro Two—house— ver."

Hoping Trung Dung or Captain Wehr was trying to reach me, I tried to establish contact with whoever was calling.

"Station calling Bunkhouse Zero Two, I am unable to read your transmission. Over."

Unfortunately, enough of the response, which was again mixed with interference, could be understood to make the message clear, "—ro Two—Intruder. Still unable—fly."

It was Trung Dung telling us that our ride home was still on the ground. I responded to let them know we had received the transmission and understood the situation. "Roger. Understand Intruder unable to fly. Zero Two, standing by. Out."

Trung Dung's Radio Room

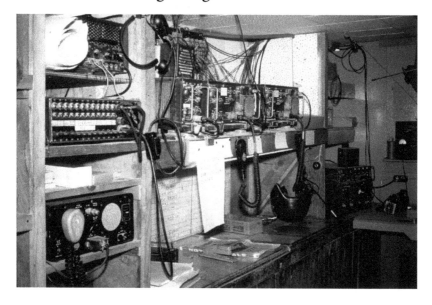

This is the radio room with which I was attempting to communicate from our LZ. Above, is one side of Trung Dung's radio room that was filled with primary and backup communications equipment. And, we had an antenna tower that was approximately forty feet tall to send and receive radio signals. So, it was unusual for us not to be able to communicate.

On the next page are three of our dedicated communications specialists. I knew I was communicating with one of them but couldn't tell which.

A-502 Communications Specialists

Above, Sergeant Jean Lavaud on his shift monitoring A-502's radio network. This was Jean's world and he knew it well.

Corporal Bob Hawley (L) being promoted to sergeant and Specialist James Miller (R), two of the other men who manned the radio room during shifts that covered every twenty-four-hour day. If something occurred on their watch, they immediately woke those responsible for responding to the emergency and alerted support units as directed.

My response was repeated a couple of times in case they were having the same difficulty with the reception that we were.

Well, I thought, dropping the handset on the ground next to the radio— *We're going to be spending the night away from home.*

So be it, I thought. There had been many nights out before. While unique, this one would simply be added to my growing list of Vietnam experiences.

Other overnighters had always been planned or, at least, the possibility anticipated. And, on the others, I was typically within the range of support units that could provide an arsenal of weaponry. More often than not, I was also in an area where reinforcements could reach my unit. In the past, communication was essentially taken for granted. I was rarely in an area where communication wasn't possible, but on this occasion, it seemed everything had gone to the devil. I had a very small team to be where I was, had civilians with me, had no support, and was in communication with—absolutely no one. Our situation in the distant and very remote valley had changed, and we were very much on our own.

Without the civilian presence, our situation wouldn't be much of an issue at all. We would simply move down into the jungle and wait out the weather. If the rain didn't let up, as was sometimes the case, we would simply take on the role of a long-range patrol and walk back to Trung Dung.

A month or so earlier, I had been out on a ten-day patrol and had returned with a short beard. While this would be a much shorter trip, it would be very difficult for the civilian media—who didn't seem inclined toward such an adventure.

With some feelings of regret for not making sure Culhane and his crew had gone back with the others, I began to feel very responsible for the three-man media team. I didn't want them to have to walk the very long distance back to Trung Dung and didn't want to think about what might happen if we encountered an enemy unit along the way.

From the mountainside, I looked out over the still thickening black sky. With the cloud cover and rain as it was, darkness was probably less than an hour and a half away. It was becoming more apparent that we were going to have to move off the LZ.

I had taken some comfort in the thought that the enemy didn't know how few of us remained, so they may be reluctant to attack. But, if they did eventually discover our number, adequately defending the LZ with our small unit would be difficult. And, when we did leave the LZ, we certainly didn't need to be wandering around in the jungle on Charlie's turf after dark. So, after talking to Sergeant Nguyen, we agreed that if we moved before dark, we could find a relatively safe and defensible location to consolidate our troops for the night down in the jungle.

As we stood in the rain and prepared to move our small unit, I pondered about how significantly our situation had changed since morning. We had come with the morning light as rescuers, but as darkness drew near—we were the ones in need of rescue.

Turning to go back toward the radio, I thought about the unopened box and letters on my bunk. The box and one of the letters were from my family. It was a certainty that the box contained a selection of good things to eat. The last one they sent had been filled with chocolate chip cookies baked by my mother. They were broken into a hundred pieces when they arrived, but that didn't seem to bother the Montagnard children who had helped me empty the box.

The letters and box back at camp were still unopened because I had been so busy with the Montagnards there hadn't been time to open them.

Funny, I thought, *the Montagnards are closer to the box and letters than I am and we are now where they had been . . . and virtually in the same situation.*

Once more I reached down and picked up the radio handset.

"Bunkhouse, Bunkhouse, this is Bunkhouse Zero Two. Over."

The interference was still very bad, but an answer was audible.

"Bunk—Go."

"Roger, Bunkhouse. Ahh . . . our situation here will be untenable after dark. So, we are preparing to move. Over."

Bunkhouse's next transmission was still very scratchy and weak, but at least all of it could be understood and it sounded like Bill Lane.

"Roger, Zero Two, understand. Wait one, while we check with Intruder. Over."

A couple of minutes passed, then Bunkhouse called back.

"Zero Two, Intruder One would like you to stand by a little longer. He says the weather appears to be breaking in there. Over."

"Roger, that's good news, but let him know that our situation becomes more dangerous the longer we wait. Over."

"Roger, Zero Two. If you haven't heard from us in thirty minutes, make your move and we'll contact you for new coordinates in the morning. Over."

"Roger, Bunkhouse. We'll be standing by. Zero Two—out."

Culhane was walking over to ask me about the choppers and had overheard the last part of my transmission to Bunkhouse.

"Are they on the way?" he asked.

"Well, the weather in Nha Trang is breaking, so they may be able to get off the ground soon."

Looking back out at the sky, I'm not sure I believed the Bandits would ever get off the ground. Indeed, I was fairly convinced that they wouldn't and thought about calling everyone in and heading downhill into the jungle. But, if Wehr was going to try to fly in that weather, we could wait another thirty minutes. An extraction before dark was certainly the desired conclusion to our mission. Even though I had only known John Wehr for a very short while, he seemed to be the kind of man who would fly if he could.

Culhane had walked back over to update his crew when Sullivan came up to where Trung Si Nguyen and I were standing.

"What's our situation?" he asked.

"Well, if they can make it in the next thirty minutes, we'll fly back. If they can't, we'll secure a place in the jungle tonight, then move to another pickup point in the morning."

"A ride today would be nice," he said.

"You're right about that. We're all soaking wet and I'm getting cold. Who would believe you could get cold here?"

Turning to once again survey the clouds to the east and north, one of the two directions from which Intruder One would approach if he was airborne, I said, "Look at that stuff. I don't know how he'll get through it."

From a rock I found to sit down on, I watched as the rain over the LZ diminished to a cold drizzle and the ceiling overhead began to lift. But as the minute hand on my watch moved from marker to marker, it appeared that there would be no significant improvement in the weather before dark.

Getting up off the rock, I slapped my rain-soaked hat against my leg and repositioned it on my head. I shook the rain off my M-16 and told Ahat to put his radio backpack on. It was time to move.

Walking over to Trung Si Nguyen, who had just come up from our perimeter, I asked if any of his men had reported any enemy activity. He said that one of his men on the western perimeter had seen movement and said he could hear people talking, but couldn't see anything clear enough to fire. Earlier, when it began to appear that we would be out here overnight, Trung si had told his men that he didn't want them wasting ammunition. So, he said that was the reason the man hadn't fired, he had no target.

The fact that the detected movement was on the west side of the ridge was good news for me. When we moved out, we would be heading down off the ridge to the northeast. Hopefully, the enemy wouldn't be in front of us, but since they knew the area very well, they could turn up almost anywhere.

Trung Si Nguyen wasn't any more excited about moving down into the jungle than the rest of us. He had grown up west of Dien Khanh and knew that area very well, but he was unfamiliar with the mountain area where we were then located. He felt uncomfortable about our situation and told me he didn't like being in a place the enemy knew better than he did.

Trung si knew we couldn't wait on Intruder any longer. It had already been thirty minutes since my last transmission with Trung Dung.

We picked up our gear, called our perimeter security, and prepared to move down over the hill. I made one last call to check in with Trung Dung, "Bunkhouse, Bunkhouse . . . this is Bunkhouse Zero Two. Do you read me, over?"

This time there was no response at all. So, in the blind, I sent a message that we were moving off the LZ and down into the jungle. "In the blind" simply means that I was sending a message even though I didn't know if anyone could hear it.

With that, we headed down toward the jungle. We had only gone about fifty yards when the radio on Ahat's back began to crackle and I thought I could hear my callsign. I gave the signal to stop. Then, the radio seemed to boom with John Wehr's voice.

"This is Intruder One, Intruder One . . . looking for Bunkhouse Zero Two, Zero Two. Over."

His transmission was still a little scratchy, but it was loud and clear enough that I knew he was up there somewhere. And—he was close!

There were spontaneous cheers of "Okay!" and "All right!" from the CBS crew as we stood at the edge of the jungle.

I responded immediately, "Roger, Intruder One. This is Zero Two. Where are you? Over."

"This is Intruder One. I'm not sure. We're still in the rain. We've been using the river as a reference and feel like we are near and to the north of you. Can you see us? Over."

"He's out there somewhere," I told everyone. "Look for him."

There was a sea of clouds over the valley. I scanned the swirling mass from west to east, focusing at various distances hoping to see him. Nothing—I couldn't see anything but clouds.

"Does anybody see him?" I asked.

There was an assortment of negative responses. Then, off in the northeast at some distance, I thought I saw something moving between the clouds. Looking farther ahead for what I hoped was Wehr, I waited to see if he would fly across my line of sight.

What happened next was extraordinary. Hollywood would have a difficult time recreating what I saw as dramatically as it occurred.

While I'm reasonably certain it was caused by the reflection of light from the setting sun penetrating the clouds, I have never disregarded the possibility of divine intervention. But, through a tiny opening in the clouds directly in the center of my line of sight, there was a sudden bright, swirling, starburst flash of light that was visible—for only a split second. The flash was as bright as any lighthouse beacon I've ever seen. Even if it were the reflection of sunlight on the wet Plexiglas nose assembly on Bandit's helicopter, the occurrence was remarkable!

"I've got him!" I shouted to the others.

Then, quickly, I called Wehr.

"I've got you, Intruder! Break southwest. We're at your ten o'clock position. Over."

"Roger. Breaking. We're coming around now," he answered.

Then, alerting Trung Si, "Call our point men back quickly. We didi!" Trung Si had sent two men ahead to scout a trail.

"Yeah, good, Trung Uy. I tell them now."

With that, I took the only smoke grenade I had with me, a red one, from my web gear, pulled the pin, and threw it up the hill behind us.

"Intruder, we've got smoke out. You should have no problem spotting us when you break through the clouds. Over."

"Roger, the clouds are thinning now."

There was a short pause. Then, Captain John Wehr, leading

his rescue team, burst out of the clouds. "Roger, I've gotcha. I assume all that red smoke belongs to you, Zero Two. Over."

"Roger, it does. We didn't want you to miss us. Over."

"Not a chance. Get ready. We're coming down and don't want to be there any longer than we have to. Over."

"Roger, no problem. We're ready. Come get us."

We quickly moved back up to the LZ and created a hasty perimeter to provide security for the incoming choppers.

To determine how we would load onto each helicopter I radioed Wehr for a count on the slicks.

"Intruder, Zero Two. Over."

"This is Intruder. Go, Two."

"Roger, Intruder. What's your count on slicks? Over."

"Our count is two. Over."

"Roger, Intruder. Zero Two, out."

As the flight cleared the clouds, we could see the second slick and a lone Wolf Pack gunship piloted by, who else, Dolloff and Black. Dolloff's ship was moving from the rear of the formation to take the lead.

With only two slicks, I organized and divided the seventeen of us between the two choppers. There needed to be no confusion about who was getting on which one when they landed and there needed to be no delay in loading.

The load for a Huey was normally about six to eight Americans with full combat gear. However, I had once seen one take off effortlessly with twelve Vietnamese troops on board. Because of their smaller size, there was less weight to lift.

I told Sullivan to take Culhane and his crew to the lead chopper as soon as it touched down. Then, since we hadn't brought a lot of gear, I told Trung Si to take all of his men to the second chopper.

As our rescuers grew near, I asked Culhane, "Are you all ready?" "Yes, we are," he said. "And, Lieutenant, thank you for getting us out of here."

"I'm not the person to thank." Pointing toward the incoming choppers, I said, "Those are the guys to thank. And, when they get here, I'd like you and your crew to get on the first one down, okay?"

"Sure, we'll be glad to. We're ready to get out of here," he said.

They all seemed relieved that we wouldn't be going into the jungle. Still soaking wet and cold, I know I was glad to be flying rather than walking.

"We'll talk when we get back at Trung Dung," I said, and added, shaking his hand, "Thank you for coming with us."

When I looked back to see where the choppers were, the Wolf Pack gunship appeared menacing and seemed to approach with attitude as it swooped directly over us at less than fifty feet. He was low, very low.

Then immediately behind Dolloff and Black came Captain Wehr and the other Bandit slick. The two choppers touched down at about the same time and I gave the signal to load. I watched as everyone scrambled aboard and I pushed Ahat, who was still at my side—as he always was, toward the lead chopper.

When the last man boarded the second chopper, I jumped on Wehr's ship, 113, next to Ahat, and yelled, "Go!"

Then, we quickly lifted off the cornfield and rotated to the northeast. As we departed, we could hear the crack of gunfire—the NVA were taking their last shots at us. Captain Wehr flew down between ridges and out into the valley toward the river.

On the way back to Trung Dung, our sky route took us through and around many clouds. As we flew, I thought about how lucky we had been. During our attempt to help Mang Quang and the other Montagnards, many things could have happened that would have yielded very different results. With that in mind, as we passed through the heart of one particularly thick cloud, I am not embarrassed to say that I whispered a brief prayer of thanks that we were all returning safely.

When we approached Trung Dung for landing, I was surprised to see several trucks and reinforcement units just leaving the runway. The troops should have long been released and I didn't know why trucks would be there. I didn't understand, but I would later learn why they were all still there.

After we settled in and landed at Trung Dung, I waited for Culhane and his crew to unload, then moved to the spot between Captain Wehr and his copilot.

"Well, I guess you know what you did was way above and way beyond," I quipped.

He laughed.

"Well, we decided we weren't going to leave you and your team out there. We got all the Montagnards, so we decided we'd get you too."

"No, truly . . . I can't believe what you all did." Then, I added, "I feel certain you saved our lives. We were a long way from home, and you and I both know you didn't have to come. 'Thank you' seems like so little to offer you."

We shook hands and just before I turned to climb out of the chopper, I offered an observation, "Birds don't even fly in weather like that. We owe you one."

"No," he said. "You owe us nothing. We were proud to have been a part of the mission. Just take good care of the villagers for us."

"It'll be a pleasure," I said.

Then, after jumping down onto the Trung Dung runway, I walked to Wehr's window with one last request and comment. "Please thank everyone at the 281st who was involved in the mission for their tremendous effort. We could never have done what we did without you . . . you made it happen," was my final remark.

The two choppers were at full power and were lifting off when I reached the small rise at the entrance to the airstrip. I stopped and turned around to watch them leave. As the two ships flew over the south wall, one behind the other, I thought about what remarkable men occupied those two helicopters with the numbers "281" painted on their nose and side doors. And, I thought about all the others that had flown to the village without knowing what dangers they might face. Individually; they were the Bandits, the Rat Pack, and the Wolf Pack. Together, they were—the Intruders.

CHAPTER 21

Many Years Later

AFTER WALKING AWAY FROM the side window of John Wehr's ship on the last day of the rescue, I wouldn't see him again for nearly fifty years.

Now, flash forward nearly a half-century, when I was invited to speak at the Airborne & Special Operations Museum in Fayetteville, North Carolina, the home of Fort Bragg. The occasion was to dedicate a memorial to the 281st Assault Helicopter Company. The memorial was to recognize the 281st as the U.S. Army's first Special Operations Helicopter unit. This was a tremendous honor and credited the 281st for being the forerunner of modern-day Heliborne Special Operations.

After I had spoken and as pictures were being taken, a man walked out of the crowd and extended his hand. He said, "I enjoyed the things you said about our unit."

The man looked familiar. Taking his hand, I said, "Thank you. They are all true."

"I know they are," he said. "My name is John Wehr"—it was "Bandit Leader," "Intruder One."

Spontaneously, I squeezed his hand, saying, "It's been a very long time, but I am still extremely grateful for your courage and what you did."

Then—two old warriors hugged.

Shutters began snapping anew and people asked us to turn around for pictures, so we did.

Our reunion was not unemotional. Even though we hadn't spoken or seen each other in so many years, it was as if not a single day had passed since we were last together. We shared a unique and amazing bond that often exists between veterans.

My family knows what John did for me, so he will always be very special to all of us.

When we were stranded on the cornfield, I wondered how the helicopters would be able to reach us in such terrible weather. Well—also many years later, I finally learned the answer.

During a 281st reunion, I was enjoying an adult beverage with Jay Hays. Jay served as the Crew Chief on John's helicopter, Bandit 113. I have mentioned that ship previously, it was my command ship the first day of the rescue and it was the one used to extract me and half of my stranded team the last day of the rescue. So, while a piece of machinery, it holds special meaning for me. And, Jay Hays knew every nut and bolt on 113.

Shortly after the first edition of my book was published in 2004, Jay called me and introduced himself. Jay was living and, as of this writing, still lives in Lordstown, Ohio. The very next Christmas after our introduction, he turned up at my home in Atlanta, Georgia, dressed as Santa Claus. And, he was sitting in a full-size replica of Santa's sleigh. Remarkably, a Georgia State

Trooper had helped him find our house. But, that's another story. Suffice it to say that, throughout the intervening years, Jay and I have become close friends.

Anyway, while we were enjoying our adult beverages, I casually mentioned to Jay that I didn't know how they made it back to us through all the nasty weather.

"I can tell you if you'd really like to know," he replied.

When I assured him that I did, the following is, more or less, the story he shared with me.

After dropping Montagnards off at Trung Dung, Bandit 113 was in line to be refueled in Nha Trang when a storm blew in off the South China Sea. Jay said they could barely see the sky, much less fly in it.

While waiting for the weather to clear, Captain Wehr continued to communicate with Trung Dung. During one of those communications, he learned how few of us were still out on the cornfield. Deciding that the danger was too great to leave us out there and against flight regulations, John took off with two other crews that volunteered to follow him back to the village. One ship was a Bandit troop carrier (slick) and the other was a Wolf Pack gunship.

Jay said that, once airborne, the flight of three choppers followed the Song Cai River west because they knew it would take them somewhere near the village and the cornfield where my team and I were stranded. He said that they were flying in a sky full of clouds that got so thick they couldn't see the ground.

When Wehr felt they were near, he tried to reach me by radio but had no luck. He decided he needed to try to get low enough to see the ground so he could figure out where they were. Not wanting to risk the entire flight because the area was mountainous, he directed the two trailing choppers to maintain altitude while he started down.

Captain Wehr radioed the other two choppers that he would continuously report altitude and airspeed as he descended. He told them that if his transmissions suddenly stopped, they would know that he had hit a mountain. They should then gain altitude and find another way down.

When mentioning them during speaking engagements, I have said many times that the men of the 281st Assault Helicopter Company were a very special breed of man—and they still are.

During the same reunion where Jay shared his information, another pilot told me something I didn't know.

When the rescue was underway, the pilots and crews of the 281st lift helicopters had created a contest. The winner of the contest would be determined by the crew that could extract the most Montagnards in a single load. The count was reported as "souls on board." The winner of the contest was the crew that lifted off with twenty-two souls on board, an assortment of men, women, and children they had freed from slavery and abuse. In my book, and this is my book—all the crews were winners!

After our reunion, John Wehr and I, who were about the same age, became more than fellow veterans, we developed a friendship. We communicated by both phone and email and I learned more about him personally. I discovered that despite the aggressive demeanor of a fierce warrior he had displayed in Vietnam, and although he had been awarded several awards for heroism, John was, in reality, a very caring and humble man. I used the word "was" because, sadly, John died on December 4, 2018, at age 74. At least, this warrior died peacefully.

It is a certainty that many, particularly his family, will never forget John R. Wehr of Augusta, Georgia. I know that my family and I will not forget the man who used the callsigns "Bandit Leader" and "Intruder One" because, as you will read in my "Final Notes," I have little doubt that he and those Intruders who came with him saved my life and those of the men who were with me.

Nearly 50 Years Later

John Wehr (L), Jay Hays (C), and me (R). After our spontaneous reunion, several people were calling to us for a picture, so that's the reason we are all looking in different directions. This picture was taken by my wife, Amy.

The last time we were together, we were young, in Vietnam, and we were warriors. Now, we were all loving grandfathers and—very glad that we were.

The Memorial

It was the installation of this memorial that reunited me with John Wehr and Jay Hays, two of the men who risked their lives to save mine and those of my team when we became stranded in the Valley of the Tigers.

CHAPTER 22

De Oppresso Liber

OUR TEAM HAD DONE what we set out to do. When the mission was complete, we had freed one hundred sixty-five Montagnard villagers from a life of slavery and abuse. And, in the process, the members of A-502 had lived the Special Forces motto and freed the oppressed. For me, De Oppresso Liber was more than a motto. If you wear the uniform of an American Special Forces soldier—it is a promise.

In carrying out the mission, we had been very fortunate and maybe just a little bit lucky. But that fortune and luck had allowed us to rescue defenseless families, among them was a woman in her eighties and the youngest was a baby who had been born the day before we arrived.

Equally significant as the lives saved, not a single member of the rescue team had been wounded or killed. My greatest concern when committing to the rescue attempt was the number and seriousness of casualties that might be incurred by our rescue team. But, in the end, the rescued and their rescuers left the village safe and sound, something for which I have always been very grateful.

On the following pages, you will see some of the things I saw on the last day of the rescue. The first picture is of the man who sought help for his village and started it all.

Mang Quang — Dropped Off

Here, a Bandit crew chief helps Mang Quang as he jumps out of the hovering helicopter on the third and final day of the rescue.

Mang's family was safe back at Trung Dung and he could have been there with them. Instead, he insisted on going with me to help in the recovery of the remaining villagers.

He stayed at my side all day except to work the loudspeaker and bullhorn. Then, when villagers began to arrive, he went to greet and organize them for the trip to Trung Dung and freedom.

They looked funny but, by the end of our time together, we had developed a few very basic hand signals by which we communicated. When we saw the villagers approaching, he looked at me, pointed to himself, then pointed toward the incoming group. I nodded and motioned him to go. He may have been primitive, but believed him to be very intelligent—I still smile when I think about him.

Telling the Story

David Culhane, his cameraman and sound man, Don Tate, and the other media representatives were extremely courageous to go with us. I've said it before, but we were, quite literally, in the middle of nowhere. They were all professional and all determined to get their individual stories. David and his crew were maybe a little too determined. If it had been up to me, to ensure their safety, I would have left all of them at Trung Dung. But 5th Group had given them clearance to be as involved as they chose to be and they all chose to go. Remarkable, because they were all—unarmed.

Above, you can see the female reporter armed only with a pad, pencil, and a camera. To her right is John Daly, assigned guard.

Ultimately, it was the members of the media who documented the story of the rescue. The picture on the previous page, this one, and the next three pages were provided courtesy of the CBS Television News Archives and have been used with great appreciation. It is with considerable gratitude that I thank David Culhane and all the others for their commitment to telling the story of what happened in the Valley of the Tigers.

Villagers Arriving — on the 3rd and Final Day

This group of women and children arrived smiling and happy to be safe.

Villagers — Ready for their Flight to Freedom

Above: Mang Quang (Center), organizing villagers for departure and ride back to Trai Trung Dung.

Below: American advisers help women and children to a hovering chopper. Surprisingly, the villagers didn't hesitate to board.

Families with Everything They Could Carry

Families arrived carrying all that they could and those things most precious to them. The mountain was steep and the trek difficult.

Some of these families had scattered after our first two visits. As a result, they had to walk almost three hours to reach the cornfield.

Going Where the Job Took Them

The last time I saw the woman I have referred to as "the female reporter" (left foreground) she was boarding one of the Bandit choppers. I am sorry to have forgotten her name, she deserves recognition by name. When it came time to leave, she insisted on waiting for all the Montagnard families to be safely loaded for their trip to freedom before she would board—she loaded with the last families out.

In Vietnam and other war zones, members of the media willingly and eagerly go where their jobs take them. Often, they find themselves on the front lines or, as the young woman pictured above, in enemy territory. As a result, they risk the same fates that can befall the troops who are the subjects of their reporting.

Another female war correspondent was killed in Vietnam as she went about her job. Her Name was Dickey Chapelle and she was killed when a soldier in front of her triggered a booby trap.

My concern for the female reporter was genuine as were the dangers we faced. Ignoring my personal feelings about her presence, she convinced me that I needed to let her do her job—and I did. Still, now and then, I looked around to see where she was.

UPI Radiophoto

Going Where the Grass Is Greener

A Montagnard family makes its way through high grass to waiting helicopters for evacuation from a Red prison farm near Nha Trang. They were among 100 Montagnards freed by Allied forces. The Communists had used the tribesmen as slave labor. (UPI Radiophoto)

The newspaper clipping above was sent to me years ago. The picture is grainy and there is no credit, but I believe it was taken by the female reporter. Below, the CBS cameraman photographed the same family.

CBS News War Correspondent, David Culhane, in "The Valley of the Tigers"

A few days after our final trip to the Montagnard village, news of the rescue began to break in various media forms in the United States. The story appeared in several newspapers and Walter Cronkite introduced David Culhane's report on his National TV news program, the CBS Evening News with Walter Cronkite.

Culhane closed his report on the rescue with what I thought was a wonderful remark and a tribute to all those who had risked their lives to free the inhabitants of a primitive mountain village. David closed by saying—

"This is a rare occurrence in this war, an act designed to give life and freedom in a place and time noted mainly for death and destruction."

You can see David Culhane's report as it was filmed on our cornfield LZ and as it appeared on the Evening News with Walter Cronkite by going to YouTube and typing "Rescue in the Valley of the Tigers"—many of the still pictures you have just seen will come to life.

Republic of South Vietnam — 1968

ADVANCES IN COMMUNICATIONS TECHNOLOGY brought television pictures of the Vietnam War into America's living rooms. Advances in photographic technology miniaturized cameras to pocket size. As a result of this technology, I began carrying a small camera with me on parachute training jumps at Fort Bragg.

My first effort at airborne photography was an attempt to capture the opening of my own parachute. Unfortunately, I made this attempt during a training jump from a Lockheed C-141 Starlifter, a huge Air Force jet powered by four Pratt & Whitney TF33-P-7 turbofan engines.

Because the C-141 was a jet, its airspeed at jump time was necessarily faster than that of the propeller-driven jump aircraft we were used to using. As a result, the sudden and abrupt jerk caused by the rapid opening of my parachute ripped the camera from my hands. I am confident that a hunter or naturalist will one day find that camera in the North Carolina forest where it fell.

While in Vietnam, as a function of my role as S2 and S3, I often carried a camera in my pocket to document various missions and key terrain features in A-502's area of operation. After Sergeant Gordon Gilmore, a team member, built a darkroom in one corner of the team's motor pool, he taught me how to develop my film and print pictures. So, I began taking more—some of those photographs appear on pages throughout the book. Others were taken and contributed by my friend, 1st Lieutenant Bill Phalen.

After learning the skill, I taught Phalen what I had learned from Sergeant Gilmore and, in the process, I created a monster—Bill took and developed hundreds of pictures while he was in Vietnam.

I wouldn't discover until years later that one morning during the rescue, while I was focused on the rescue, Bill boarded one of the Bandit choppers and flew out with us to document the mission. The picture of our choppers disappearing into the mist was his.

Other pictures were shared with me by a teammate, Sergeant Jean Lavaud, a key member of our communication team, and others as noted. I offer my thanks to all the contributors because—nothing helps tell a story like a picture.

Compound Center — Trai Trung Dung

Home of Special Forces Detachment A-502's Base Camp.
The old cannon was discovered in the moat that surrounded the fort.

A-502 Team House

This is where Daily dropped me off the day I arrived at A-502 and where I met Thieu ta. The team house is on the left and the "Hospital" is on the right.

S2 & S3 Offices — Opposite the Cannon

The S2 Office, with Sergeant Paul Koch as staff, was on the left side and the S3 Office, with Sergeant Roy King as staff, was on the right side. When I wasn't in the field, I worked back and forth between both places. Work on the rescue involved long hours in both offices.

Fort Perimeter Moat

This is part of the moat that surrounded the fort. In days of old, attackers would have to swim or use boats to cross the moat, neither a wise option.

Emergency Bunker

If the camp were overrun by enemy units, team members could move to the bunker and continue to direct support units from underground. The bunker had maps, radios, and supplies needed for the emergency command center.

"Fire Arrow"

If the fort wall were breached, the Fire Arrow on top of the bunker could be turned from inside to indicate the direction of an enemy attack to Air Force FACs flying overhead. As you can see, the arrow had lights for a night attack. This made it easier for FACs to direct fire accurately.

Gun Jeep

My Corner of the Team House

Everyone did what they could to make their private area seem just a little less military. The cards were on the wall over my bed so that I could see them. They reminded me of some of those back home.

What you can't see is the shelf with hand grenades, smoke grenades, illumination flares, my M-16, and combat web-gear.

Fighting Positions on the Fort Wall

These concrete fighting positions were constructed around the top of the fort wall in Trung Dung's area of responsibility.

Doing What We Did

A-502 advisers overseeing training for Thieu ta Ngoc's CIDG soldiers.

Wall Top Bunker

This bunker was at the top of one of the fort's star points, so it had a wide field of view and could fire down two sides of the wall.

Flares Over a Battlefield

One of our units ambushed an enemy patrol very near Trung Dung and they fired flares to illuminate the battlefield, a too common sight.

Some of Those Mentioned

Major Ngoc Nguyen

1st Lt. Bill Lane

Vietnamese Camp CO

Team Executive Officer

Major Will Lee

Teammates

My friend, warrior, and camera buff, 1st Lt. Bill Phalen (L) and me (R).

A-502 CO

Sergeant Paul Koch

Sergeant Roy King

S2 (Intelligence) Sergeant

S3 (Operations) Sergeant

1st Lt. Tom Ross

1st Lt. Mike Sullivan

S2 / S3 Officer

Company Advisor

In Combat Gear and Ready for any Mission

Vietnamese Camp Commander, Major Nguyen Quang Ngoc was a fearless warrior, dedicated to the freedom of his country.

CIDG Troops in Formation

Some of Thieu ta Ngoc's troops in formation. These were the men we helped train and advised in the field. When we went to the field with them, there were typically only two (American advisers) with them.

Montagnard Village Dwellings

Typical Montagnard common house. This one was in Ninh Hoa.

The dwellings in the Valley of the Tigers were similar to this one, but larger. Though larger, they were well hidden by the jungle.

281 Assault Helicopter Company in Action

As important as it was, the 281st did far more than support our rescue mission. They supported A-502 and others on a daily basis. Above: A 281st ship is delivering supplies to our Dong Mo Mountain outpost. The landing area was very small and very tight.

Going Where They Shouldn't Go

The 281st often flew into places where they probably shouldn't have gone. I once saw them fly down into the jungle to deliver water through a hole that was about the same width as the chopper's spinning blades.

Men of the 281st – They Made it Happen

Major John "Jack" Mayhew, CO of the 281st while I was there, and his XO, Captain Bob Moberg.

*Captain John Wehr
"Bandit Leader"
"Intruder One"*

*Captain Glen Suber
"Rat Pack Leader"*

*Captain Bain Black
Wolf Pack Pilot-Gunner*

*Captain Ted Dolloff
"Wolf Pack Leader"*

CHAPTER 23

Day Not Over Yet

ON THE WAY BACK to the team house after watching John Wehr and the other Bandit ship leave, I stopped at the Montagnard camp, as I had many times during the past few days. Mang Quang, his wife, and his children were sitting on a bench just inside the tent. They were laughing and trying to throw small rocks into an empty C-Ration can. Mang's son was tiny, but he tried to play.

A couple of days earlier, I had asked one of the interpreters to teach me a few key phrases that could be used when speaking to the children. So, when close enough for them to hear me, my greeting was something roughly approximate to, "Hello, little ones," in Montagnard.

The reaction from the children was probably much the same as any child's, regardless of their nationality or language used. They became very quiet and began acting very bashful.

With some prompting from his father, Mang's small daughter, who looked to be near four or five years old, responded with her equivalent of "hello." Then, much as a shy child would from anywhere in the world, she nuzzled close to her mother, who was on the bench next to Mang Quang.

Then, Mang, with one arm on his small son's leg, reached for his daughter with the other. Gently, he pulled her close and held her snugly. After her father whispered something to her, she showed me a tiny, but wonderful smile, which I returned.

About to leave, I was taken by surprise when Mang Quang stood, reached out, and again took one of my hands in both of his, just as he had when we first met. Then, surprisingly, he said, "Thank you, Trung Uy." And, then, he bowed slowly—he had obviously asked for a language lesson of his own.

His unexpected simple gesture of thanks was deeply touching to me. Responding, I smiled and returned his bow. I don't mind telling you that it was all I could do to maintain my composure. For all we had done, all we had experienced, and everything we risked, Mang Quang's near whispered, "Thank you" was enough for me. I smiled once more, nodded, and headed for the team house to get out of my damp uniform.

As I walked away, I looked back over my shoulder to Mang Quang and his family, then beyond to some of the other families and individuals who were now safe at Trung Dung. Among them, a baby girl who had been born the day before we arrived. Not far away from her was a woman well into her eighties. It was a warm and gratifying sight for which only Mother Nature could have provided an appropriate backdrop—and she did. In the west, beyond the villagers, deep golden shafts of light from the setting sun pierced thinning clouds and caused the horizon to glow. It was a beautiful and fitting close to a day that couldn't have been much brighter for the men of Special Forces Detachment A-502—or me. I had been given a special opportunity to fulfill a promise made to three mountain villagers by simply carrying out the Special Forces motto, *De Oppresso Liber,* free the oppressed.

Turning back, I took my floppy soggy field hat off and stuck it in my web gear. Then, reaching down into one of the leg pockets on my fatigue pants, I pulled out my rolled beret, which was only slightly drier, and positioned it on my head. While I couldn't wait to change, and even though this one was still wet, I was never prouder to be wearing the uniform of an American soldier.

Then as I continued to walk toward the team house, I almost ran into someone who suddenly stepped out of the darkening shadows. It was Thieu ta. Trung si had radioed him that we were on the way back.

Talking as we walked, "So, you did it, Trung uy! You got all the people!"

"No, *we* did it, Thieu ta."

"The tiger patch I gave you still has no bullet holes?" he asked, laughing as he spoke.

"No, it has no holes I am very glad to say."

Then, Thieu ta quieted and became more serious.

You know I wasn't going to leave you out there, right?"

"What do you mean?"

"When I was told that the helicopters couldn't come after you, I had the reinforcements gather at the runway and I called for trucks."

"So, that's why they were still out there?"

"Yes, I wasn't going to wait until morning. Two of my Vietnamese Special Forces soldiers who were on the ground out there during your second mission said they could take me to you. So, if the helicopters didn't come for you, we were going to drive the trucks as far as we could, then begin walking."

Thieu ta then showed me a map that Sergeant King had made and given to him for the trip. There was a circle around our location on the LZ. While it would have taken him some time to reach me, that's what he intended to do. Is there any question why Nguyen Quang Ngoc and I shared a lifetime friendship?

When we reached the team house, I said goodnight. Ngoc slapped me on the back and said, "You made it!" Before thinking, I said, "Yes, I survived the day."

Then, just as he had the very first time we met, waving his finger, Ngoc smiled and said, "Day not over yet, Trung uy."

EPILOGUE

AGAIN ON THE ASPHALT runway at the Nha Trang Air Base with my duffle bag, my tour of duty was complete. I was waiting for the helicopter that would ferry me back to Cam Ranh Bay, where I would board a flight back to the United States—and home.

When I stood on this spot nearly a year earlier, I wondered if I would survive the experiences that awaited me and return to the same place. Since pondering that though, much had happened. My time in Vietnam had been a series of unique life experiences and I was going to leave with many new things to ponder and for which to be grateful. Most importantly, I had survived the experience.

Just as I noticed the incoming helicopter, an Air Force airman walked out to the edge of the runway where I was sitting and pointed toward the inbound chopper.

"Lieutenant, that's your shuttle on approach. Have a good trip home."

"Great. Thank you."

A collective cheer arose as the wheels of the Northwest Airlines jet lifted off the runway in Cam Ranh Bay. Everyone on board seemed very happy, some euphoric, to be headed back to the United States.

As we turned east out over the South China Sea, I looked back to watch the Vietnam coastline drift into the distance. Before the coastline was no longer visible, we were high enough that I was able see far to the west. The mountains that were home to the Montagnard village were easy to see. With a birds-eye-view of the area, it was not surprising that it appeared vast, rugged, unwelcoming, and extremely remote. But, as I continued to look out over the mountains, I felt a smile make its way across my face.

Memories of time spent in the distant jungle and on the village cornfields sprang to mind. My most vivid and poignant memory was the moment that Mang Quang first saw his family after fearing that they might have been killed. That singular and very special moment made all that I had risked during my tour of duty worthwhile. I have never been prouder to be an American soldier.

Finally, knowing how the Vietnam War ended, I would do it all over again just to experience the moment with Mang Quang in the Valley of the Tigers—one more time.

If you have read *Along the Way*, you can stop here and go to the next section, Author's Final Notes. If you haven't read it, I won't cheat you out of my homecoming, please continue reading.

Turning to look ahead along the way home, I began to think about my arrival back in Pensacola. My homecoming would be quiet— very different from the one my family had planned for me.

Near the end of my time overseas, my mother often wrote about the plans that she and my Dad were making for my homecoming. They had arranged for a band from Rosie O'Grady's, one of the local nightspots, to welcome me at the airport. I, however, didn't want a band or any other fanfare at my homecoming. There was now too much controversy over our country's involvement in Vietnam. And, I didn't need or want a parade. I had volunteered to go where I went and do what I did. All I wanted or needed was to see the faces of my family and have a quiet trip home from the airport. For those reasons, I hadn't told my parents when I would be returning, so they had no way of knowing that I was already—on my way.

After a very long flight back to the west coast of the United States, another long coast-to-coast flight loomed ahead of me. After a plane change in St. Louis, we finally turned south. It was late afternoon when we landed in Birmingham, Alabama. We were scheduled to be on the ground just long enough to drop off and pick up passengers. So, anyone who was continuing to Pensacola and beyond was asked to stay on board. When we took off again, the flight to Pensacola would take only about 30 minutes.

Knowing my mother would barely have time to get to the airport in Pensacola before I did, let alone organize a reception, I thought my family should probably know that I was near. Going to the front of the plane, I introduced myself to a flight attendant and told her that I needed off for just a few minutes to call home and I explained why. Before she could respond, a voice from the cockpit said, "Go ahead, Lieutenant, we'll wait for you. Just make it as quick as you can." One of the pilots had seen me board the plane and heard me speaking to the flight attendant.

When my mother answered the phone at home, I asked, "Do you have a place for a tired soldier to lay down?"

Taken by surprise, her voice began to quiver, "Oh, my God. Is that you, Tom? You sound so close. Where are you?"

"Birmingham," I said.

"Oh, thank God," she sighed. "But we wanted to do something special for you." Then, she started to cry.

"Mom, I have to go. The plane is ready to leave. I'll be home in about forty-five minutes to an hour. Will you call Dad and let him know I'm coming home?"

"Yes. Are you all in one piece?" she asked with a mother's concerned voice.

"Yes, I'm fine. Mom, I have to go now. I love you."

"I love you, too. I'm so glad to get you back from that terrible place."

About forty-five minutes later, the propeller-driven Eastern Airlines Silver Falcon circled in over Pensacola Bay and touched down on the airport runway. As we taxied to the terminal, I could see my mother and sister, Polly, at the gate. I had barely cleared the bottom of the steps before being rushed and smothered in their embraces— I was home!

After collecting my bags, we went to see my father at work. He was in the back of our family business and had his back to me when I walked through the front door. When he heard the door, he turned, and in an instant, his face lit up. He beamed with a huge smile and quickly surrounded me with a bearhug.

Such was my homecoming—the only one I needed or wanted. I was glad enough to have made it home and been given the chance to do both the "something good" and "something meaningful" I had hoped to do.

Valley of the Tigers — A Fitting Last Word

This is an enlarged picture of the mountains west of the My Loc outpost where Mang Quang and the other two began their journey.

After the rescue, during a description of the Montagnard trio's arduous trek, Mang Quang said that, as they hurried through the jungle, they came very near a large tiger lapping from a pool of water. He said he was afraid the big cat might attack them as they ran by, but the tiger had already seen them and there was nothing for them to do but continue on. Fortunately, the tiger only raised its head and watched as the three ran by on their way. Rubbing his belly, Mang Quang said the tiger's stomach must have already been full.

So, it seems the response given to David Culhane's question of, "Where can I say we are?" was clearly an accurate one.

The Valley was Their Home

This magnificent animal looks like one of those I saw roaming in the wild not far from the Montagnard village.

Unfortunately, the tiger population in Southeast Asia has been seriously poached to such a degree that wild tigers have not been reported in Vietnam since 1997. I hope that they simply haven't been seen and still roam in the thick jungle where the village used to be or in other remote areas.

AUTHOR'S FINAL NOTES

The following is a closing note about Nguyen Quang Ngoc, the man who gave me the support and troops needed to do the "something good" I had hoped to accomplish.

When I went to say goodbye to him before returning home, Ngoc offered his hand and gripped mine firmly, but warmly. Then, looking intently into my eyes—he thanked me for serving in his country.

Ngoc Nguyen, "Thieu ta", was the first and only person to thank me for my service for many years. Later, it became customary for Vietnam veterans to thank each other for their service and offer a "Welcome home!" Eventually, thoughtful and appreciative American civilians also took up the practice. Since I had volunteered for service in Southeast Asia, I never really expected or required thanks from anyone. Nonetheless, I must quickly add that each of the many, many times I have now been thanked for my service or welcomed home, I felt warm appreciation and a sense of pride.

Still, when I think of all the many very kind gestures of thanks offered to me, Ngoc's is the one that stands above the rest and is the most important. For me—his was enough. My presence had meant something to someone.

Sadly, my friend Ngoc died on February, 5th, 2011 in Houston, Texas. Ironically, the man who fought so fiercely for the freedom of South Vietnam was born in Son Tay, a very well-known suburb of—Hanoi.

Lifetime Friends – Ngoc and Kim Nguyen

Ngoc, Kim, and I shared several lunches or dinners during my time at Trai Trung Dung. Here, I had been out on ambush the night before and arrived at lunch after it started. When he saw me coming, Ngoc called to me, "Trung uy, come sit here, I saved you a place next to Kim." That was always a seat of honor.

Farewell with Well Deserved Honors

When he died in 2011, Ngoc was buried with the full Vietnamese honors, which he had earned and to which was entitled. I still miss my friend like a brother.

Vietnamese Special Forces LLDB Patch

This is a patch Ngoc presented to me the morning after our first mission together on Buddha Hill. Because it was a Vietnamese insignia, it wasn't sewn onto my uniform. Rather, like the tiger patch he gave me the night before the rescue, it was encased in clear plastic and was made with a buttonhole attachment at the top so it would fit on the pocket button of my jungle fatigues. Over fifty years later—I still have it, a memento of immeasurable significance to me.

This final note is a personal one that simply demonstrates just how tenuous our presence here on God's green earth might be. After serving in Vietnam, I have never taken my existence for granted.

About a month or so after the last day of the rescue, one of our night ambushes captured an NVA soldier near the northern base of the Dong Bo Mountains. During his interrogation by a Vietnamese intelligence team, he revealed that he was a member of Regiment 18B's 7th Battalion, small by US standards. But it was the large

NVA unit we were concerned might arrive at the village while we were there. The 7th was attempting to reestablish the mountain base destroyed by the Koreans.

While trying to discover when the unit had arrived and the exact route that was taken to reach the Nha Trang mountain complex, the intelligence team learned something a bit alarming to me—makes me shake my head when remembering what I was told by one of the Vietnamese intelligence officers after the interrogation of the prisoner was complete.

Ostensibly, appreciative of the medical treatment he received and to avoid potentially severe punishment, the NVA soldier shared a great deal of information. It seems the unit's route had taken it to the Montagnard village, a familiar resting point. Normally, the Montagnards would be used to help transport heavy supplies into the mountains. The soldier said that when the battalion approached the village from the west, their point unit encountered an enemy patrol (our patrol) and there was a brief firefight. Then, their point unit was attacked by American helicopters (the Wolf Pack).

The soldier told interrogators that he was with the main body of the unit when it arrived at the base of the hill leading up to the village. He said the unit commander became enraged, almost crazy when told that the villagers were gone.

Upon hearing the next part of the prisoner's account, my spine began to tingle. Continuing his story, the soldier told interrogators that while they were at the base of the mountain, they heard, then watched, as three American helicopters appeared from the clouds, landed, and then quickly departed (John Wehr and his rescue team). The prisoner continued by saying that his commander believed there might still be South Vietnamese soldiers and Americans in the village. So, he ordered his 175-200 men to deploy and begin moving up the mountain to attack anyone still there. The NVA commander was sure he had the troops to overwhelm anyone still in the village. Unfortunately for him and his troops, when they reached the village, there was nobody to be found—we were gone!

I truly don't think about it very often at all. But whenever I am reminded of what was shared with me by the Vietnamese intelligence officer, it makes me stop and think about that last day at the village, so very long ago. That day could have ended very differently. Had John Wehr and his gutsy team not taken off in a driving rainstorm and come after us, someone else might have had to tell the story of—The Rescue in The Valley of the Tigers.

Captured NVA Soldier

US advisers from Lieutenant Bill Phalen's "Blue Bandit" team were with the Vietnamese unit responsible for the capture of the NVA soldier

Here, Lieutenant Phalen is present during the hand-off of the prisoner to the Vietnamese Intelligence officer, Ky (left foreground). The prisoner was just about to be handcuffed and was surrounded by an armed squad of Thieu ta's Vietnamese Special Forces soldiers. Bill gave me this picture believing I might want it for a souvenir.

Tom – "Trung uy"

Soldier

Author

THEN

NOW

Men who Made the Mission Possible

Ngoc – "Thieu ta"

John Wehr – "Bandit Leader"

This was why I went to Vietnam

This is one of the children rescued. Now, as a grandfather,
I would do it all over again.

AFTERWORD

TO MANY PEOPLE, THE Bell Corporation's UH-1 Iroquois was just a helicopter. But, to Vietnam era veterans, the UH-1 was a "Huey" and the distinctive whop, whop, whop sound of its spinning blades often meant help had arrived.

For soldiers injured or in trouble on the ground, the sound of the Huey's pounding blades was like hearing the stirring brass call of "Charge!" from a bugle announcing the arrival of the cavalry to the rescue. A Huey's appearance over a battlefield could mean evacuation for wounded soldiers, extraction for one of our stranded Special Forces units operating deep in enemy territory, or protective air cover for an American unit in a life or death firefight. I was one of those who was more than glad to see a Huey burst from the clouds.

The UH-1 was developed and produced in response to the U.S. Army's need for a medical evacuation and utility helicopter in 1952. It first took to the air on October 20, 1956, with the original designation of HU-1. It was that designation that led to the "Huey" nickname. Even though the designation letters were later reversed to UH-1, the nickname stuck.

The Huey went into full production in March 1960 and was the first turbine-powered helicopter to enter production for the U.S. military. The single turboshaft engine provided power for the helicopter's two-bladed main rotor and tail rotor. The Huey would serve as a military workhorse for many years and I would log many hours aboard them, even parachuting from more than one.

Captain John Wehr's Huey, Bandit 113, official designation UH-1 #66-17113, served with the 281st AHC during the unit's time in Vietnam. It was one of approximately 7,000 Hueys to see service during the Vietnam War. This particular helicopter played a key role in countless missions as it was piloted and crewed by a courageous and distinguished group of men during its military service.

When NVA and VC units ambushed a supply train en route from Cam Rahn Bay to Nha Trang, the pilot and crew of Bandit 113 were immediately dispatched from a pad at the 281st's base. Within minutes of the explosion that derailed the train, 113 boldly made pass after pass over the derailment to intimidate enemy soldiers that were attempting to loot the train of its military cargo. Like a mighty guardian angel, Bandit 113 protected the train and its crew until South Vietnamese soldiers from Trung Dung and our A-502 U.S. advisers could move in and secure the area.

Because train tracks ran around the base of the Dong Bo Mountains, ambushing supply trains running between Nha Trang and Cam Rahn Bay became a pastime for NVA units based in the Dong Bos. And, as you have read, on one of those occasions, Thieu ta and I led the quick response team from Trung Dung.

In March of 1968, little more than a month after I had arrived in-country, Bandit 113 became involved in a ferocious battle with enemy units while flying in support of an American ground unit in the notorious A Shau Valley. As 113 flew in and out of the battle with its M60 machine guns blazing, it was met with a hail of incoming enemy fire marked by green tracer rounds pouring in from multiple ground locations. At different times during the intense fight, other 281st Hueys and their crews were shot down from both in front of and behind Bandit 113. On that day, Bandit 113 was the only ship of twelve to remain flyable after the fight had ended.

Crew Chief, Jays Hays, credits 113's "good fortune" to the addition of an extra M-60 machine gun to his normal singly mounted gun. With his "Twin 60s" Jay was able to pour twice as much fire onto his targets, which would certainly make the enemy want to keep their heads down.

On another occasion, Bandit 113 was on a single aircraft combat-support mission when the pilot and crew overheard a radio transmission between a ground unit and a medevac helicopter. The ground unit had suffered casualties during an ongoing battle with an enemy unit that was still believed to be in the area. Concerned that the area was not secure, the medevac pilot was uneasy about the safety of his crew. Understandably, he wanted to be assured that the landing zone was secured before attempting a pickup.

Trained and experienced at landing in "hot" LZs and knowing that the lives of injured soldiers could depend on a quick evacuation, the pilot of 113 suggested to the medevac chopper that he and his crew make the pick-up.

The men of the 281st who piloted 113 were accustomed to landing in dangerous, enemy-occupied territory because of their experience flying the extremely hazardous "Delta" missions. So, without any further conversation, the pilot of Bandit 113, diverted and immediately headed toward the embattled LZ. Upon arrival at the battlefield, the pilot moved in over the ground unit's position, landed, picked up the wounded soldiers, and flew them directly to the nearest medical facility.

Not all of Bandit 113's missions were strictly combat related. As you have also read, 113 led the way for our rescue team into the Valley of the Tigers on a humanitarian mission. That first day, Mang Quang and I were aboard 113 when Captain Wehr dove through early morning jungle mist so that we could mark the LZ and begin the rescue. After that, 113 served as the command ship for Thieu ta and me.

Then, on the last day of the rescue mission, I and my very small remaining point team, along with David Culhane and his CBS News crew, became stranded in enemy territory. It was Captain John Wehr and 113 that broke through the clouds and extracted us just minutes before we would have become engaged with a significantly superior enemy force. We would have been outnumbered by better than 10 to 1. While the "Green Beret" in me would like to think that would have been a fair fight, the realist in me knows that John Wehr, 113, and those courageous crews who flew through the cloud-filled sky with them—saved our lives. So, there should be little question why I wanted to include this Afterword and admit that I love that big green chuck of whirling, noisy metal.

After I left Vietnam at the end of 1968, 113 continued to fly pilots and crews into battle until the 281st was deactivated in early December 1970. 113 was returned to the U.S. and eventually sold to a civilian company for private use. In 2015, 113 was relocated by members of the 281st who purchased and refurbished it to its Vietnam-era condition.

Today, Bell Corporation's UH-1 Iroquois, "Bandit 113" is on display at the H.E.A.R.T.S. Veterans Museum of Texas in Huntsville, Texas. If you live near there or plan to visit Huntsville, I hope that you will stop by and see 113. You will be looking at a wonderful piece of history that defended and carried countless soldiers to safety. Then, during a couple of days in August 1968, it led the way in ferrying one hundred and sixty-five defenseless mountain villagers to freedom.

I haven't had the chance to see 113 since put on display, but I hope to one day make the trip. If you go to the museum and take the time to visit 113, I ask that, while standing beside it, you reach out and pat it on the side as if a faithful steed and say—"Tom said, 'Thank you for the rides.'"

Yes, I know—sentimental. But, I'm an old warrior now and I was given the chance to become an old warrior because of John Wehr, Bandit 113, and the other Intruders. May God bless them every one!

"Bandit 113" at the H.E.A.R.T.S. Veterans Museum

Above: Bandit 113 on its new pad at the HEARTS Museum.

Right: Crew Chief, Jays Hays, in 113 behind the twin M-60 guns he made and installed. Jay is sitting in his usual position at the left rear door that you can also see above.

Most important to me, Jay was in his usual position and with Captain John Wehr on 113 the day they flew through a storm to reach me and my stranded team in the "middle of nowhere."

ABOUT THE AUTHOR

Thomas A. (Tom) Ross was raised in Pensacola, Florida where he lived and attended college until he enlisted in the U.S. Army. Tom's commitment to both community and country was apparent at a young age when he earned the rank of Eagle Scout in the Boy Scouts of America.

In early 1966, Tom enlisted in the U.S. Army and was commissioned a 2nd Lieutenant after attending Infantry OCS. He immediately applied to and was accepted for training with the Army's Special Forces, the unit also known as the "Green Berets." After completing an intense course of unconventional warfare training, he was assigned as the Intelligence Officer of Company B, 3rd Special Forces, stationed at Fort Bragg, North Carolina.

After a year of intense training, Tom was assigned as the Intelligence and Operations Officer of Detachment A-502, 5th Special Forces in Vietnam.

During his time as a military adviser, he was privileged to witness many courageous and selfless deeds performed by both American men and women. Unexpectedly, Tom was given the rare opportunity to plan and lead a unique humanitarian rescue mission. The rescue team, comprised of Vietnamese troops and America Special Forces advisers, was ferried deep into enemy territory by the 281st Assault Helicopter Company, a daring group of men who volunteered to fly the dangerous mission, all risking their lives.

When the mission was complete, 165 mountain villagers were liberated from slavery and abuse by the enemy and given—the gift of freedom.

After returning home in late 1968, Tom joined his family's custom-design jewelry firm. Then, after serving as the company's

General Manager for several years, he sought the challenge of a larger arena with family blessings.

As his career evolved, Tom was recruited to and served as an executive with two internationally recognized jewelry companies. At one of these companies, he hosted a "Breakfast at Tiffany's" where Audrey Hepburn was an honored guest. Tom sipped tea and enjoyed warm pastries with Ms. Hepburn, a very different environment than the jungles of Southeast Asia.

In 1999, Tom and his wife, Amy, reestablished their family's jewelry firm, The Ross Jewelry Company, and they currently live in Peachtree Corners, a suburb of Atlanta, Georgia.

Veterans Monument

I am very proud to live in the community that created this beautiful monument to honor all veterans of every service branch for their service to our country, its citizens, and to the preservation of freedom throughout the world . . . past, present, and future.

As long as there are wars, there will be courageous American men and women who will step forward to protect and defend us and our country.

The Men of Special Forces Detachment A-502

1964
Day, Kenneth J.
Chastain, Joyce F.
Combs, Harold J.
Fields, Arthur
Foxworth, Louis G.
Grabey, Stanley G.
Hobby, Oscar L.
Johnston, James M.
Miner, Louis F.
Spradlin, Earnest J.
Wilson, Carl L.
Wilson, Gerry W.

1965
Batteford, Frank P. Jr.
Berg, Charles L.
Chestnut, John
Cincotti, Joseph G.
Foshee, Edgar E. Jr.
Grady, Clyde E. Jr.
Hughes, James E.
Mallare, Gerald E.
McClellan, Henry
Moore, George R.
Peters, Larry J.
Rice, Homer L.
Schreiber, Robert D.
Smeltzer, William L.
Switney, Robert
Sweeney, Robert T.
Vasquez, Jose E.

Watson, Roger

1966
Chaplin, Robert W Sr.
Charest, Robert A.
Chase, James E. Jr.
Clow, James L.
Deason, Robert L.
Gumper, Victor W.
Johnson, Dean B.
Johnston, Charles W.
Jordon, Richard W.
McKitrick, Michael L.
McMenamy, Charles W.
Rouse, Glenn R.
Salsman, Berney
Shreck, Raymond D.
Shriver, Jerry
Sturm, Henry B.
Tocci, Mark A.
Young, James W.

1967
Anderson, Roger L.
Andree, Martin E.
Arrants, Jerry C.
Ballou, Richard D.
Bardsley, Richard W. Jr.
Barnes, James M.
Blake, John H. III
Lane, William K. Jr.
Lavaud, Jean P.

Brooks
Castillo, John J.
Daly, John J.
Everett, James L.
Freedman, Lawrence
Geronime, John F. II
Gilmore, Gordon S.
Goff, Robert E. Jr.
Goodwin, Donald J.
Herbert, John
Homitz, Ronald D.
Jackson, Hugh
Jarvies, James Y.
Kentopp, James M.
Key, John C.
King, Roy
Koch, Paul L.
Lee, Wilbur L.
Madera, Edward
Miller, James H. Jr.
Morace, Albert T. Jr.
Munoz, Ferdinand
Noe, Frank R.
Puckett, Wayne R.
Reynolds, Robert W.
Sanderson
Sotello, Juan
Stewart, Mitchell G.
St. Martin, Joseph E.
Sullivan, Michael E.
Dinnel, Michael L.
Ditton, William L.

The Men of Special Forces Detachment A-502

York, Larry K.

1968
Allen, Manuel B.
Armstrong, Edwin D.
Bachelor, Hardy E. Jr.
Beeler, David E.
Brandon, John C.
Brown, Edward
Burruss, Tommy
Caldwell, Herschel E.
Campbell, Charles E.
Cheston, Elliott B. Jr.
Childs, Benjamin B.
Darragh, Shaun M.
Dawkins
Drennan, Dennis P.
Dubovick, Richard R.
Dukovic, Gary V.
Egan, Jan C.
Gray, Darrell W.
Harrell, Robert L. Jr.
Harris, Edward D.
Hawley, Robert L. Jr.
Hicks, Archibald G.
Hilliard, Sidney H. III
Hines, Robert M. Sr.
Holland, James D. Jr.
Hubbard, Lyman L.
Juncer, Dennis A.
Kerestes, Paul A.
Knorr, James R.
Land, Kenneth D.
Lane, Terry V.

McGill, Charles A.
McKay
Ochsner, Robert L.
Olt, Timothy F.
Oxenham, Randall
Palmer, Charles P.
Phalen, William C.
Phillips, William J.
Pope, Alonzo D.
Robertson, Juan P.
Ross, Thomas A.
Rupp, John N.
Sanford, H. C. Jr.
Sellers, Lawrence P.
Sheppard, Andrew D.
Sipots, Carl A.
Strong, Tully F.
Trujillo, Louis A.
Webb, Carlton E.
Weller, Richard O.
Wilson, Thomas E.

1969
Abraham, Anthony
Bemis, Donald W.
Blancarte, Edward A.
Carlson, John E.
Cooper, Gerald L.
Cottrel, George A. Jr.
Cottrel, John R.
Crabtree, Donald L.
Crockett, Charles D.
Deschamps, John T.
Dinnel, Michael L.

Downs
Eastburn, David R.
Estrada, Pedro B.
Funk, George C.
Gebhardt, John L.
Gigliotti, John L.
Guerrero, Francisco T.
Hearst, John R.
Hefferman, William F.
Hein, Charles
Hoffman, George P.
Horne, Freddy A.
Jenkins, Larry D.
Jones, James H.
Kemmer, Thomas J.
Kirby, Wickliffe B. III
Lindsey, Gene B.
Martin, John R.
McBride, John B.
McCandless, Kerry E.
Merletti, Lewis C.
Mika, Michael J.
Miller, Franklin D.
Mitchem, James W.
Olivera, Gilbert G.
Overby. Morris C.
Payne, Thomas R.
Roush, James E.
Saganella, Eugene V.
Sheridan, William E.
Short, Harlow C.
Shutley, George F. Jr.
Stucki, Gary W.
Tolbert, James E.

Lightning Source UK Ltd.
Milton Keynes UK
UKHW022007010222
398052UK00010B/2049